By 1943 German army field commanders were demanding a heavier and more powerful version of the Sturmgeschütz which had first seen combat in the French campaign. These versatile assault guns had proven themselves to be not only capable of tackling enemy pillboxes and bunkers, their original mission, but were increasingly called upon to act as tank killers. As early as February of the same year the army's planners had liaised with the Artillery Design Department at the firm of Friedrich Krupp AG regarding the possibility of mounting the 12.8cm Sturmkanone in the hull of a Tiger H3 which was then under development as a replacement for the Tiger I (1).

The unmodified 12.8cm guns were to be taken from the Pzkpfw Maus project and a specially designed gun mount was proposed as there was no longer a need to accommodate the revolving turret of the Maus. This gun was an extremely potent weapon and was capable of destroying both armoured and unarmoured targets at ranges of up to 3,000 metres. It was also capable of indirect fire and at least one company was used in this role in early 1945.

The vehicle's speed was viewed as a secondary consideration and could be sacrificed to allow for thicker armour although cross-country mobility was stressed in the original specification. From the outset it was expected that Henschel & Sohn would provide the complete hull and superstructure and that this would be based on the Tiger II but with the engine moved forward. Anticipating the problems this would cause, Henschel's designers produced what they referred to as Design A, with the engine installed at the rear, and Design B, where the engine was placed in the forward section of the hull. They successfully argued that Design A would present significantly fewer production difficulties and also provide some uniformity of parts with the Tiger II, although the hull would need to be lengthened by 30cm. The idea of a forward-mounted engine was dropped and the silhouette which most readers will be familiar with, incorporating a large box-like centrally-mounted superstructure, would dominate all future development. A full-size wooden model was later presented to Hitler who ordered that series production begin immediately.

With armour thickness ranging between 100mm and 250mm for the hull and superstructure fronts respectively and 80mm for the hull sides, the final design was the largest armoured fighting vehicle to see service during the Second World War. But it was plagued by mechanical difficulties, mostly caused by the weight of its armour and massive gun, and the hostility expressed by Henschel's engineers to almost every facet of the project is well documented. Many of these considerations are examined in the Technical Details and Modifications section which begins on page 49.

The small number of Jagdtigers produced and the dire situation that the Germans found themselves in by early 1945 go some way towards explaining the lack of photographs, as compared to earlier periods, and the reader will notice that the great majority of the images in this book are of captured or disabled tanks. This also effected the preparation of the illustrations as many of the available photographs show vehicles that have been badly damaged or burnt and in some cases I have been forced to speculate on camouflage patterns and colours. In addition many sources, including contemporary documents, are ambiguous and even contradictory and where I have been forced to draw my own conclusions I have tried to make this clear.

Notes

1. German weapon and vehicle designs underwent numerous name changes before the final production models were decided on. For our purposes the Sturmkanone would become the 12.8cm Pak 44 L/55 gun and the H3 was eventually renamed Tiger II.

At Truppenübungsplatz Arys in East Prussia in October 1943 the full-size wooden model of the Jagdtiger, and behind that, theTiger II and Jadgpanther were presented to Hitler, who is standing at far left. The vehicle in the foreground is an Italian P40 tank.

THE WESTERN FRONT, 1945

Our map shows the situation in the West from the beginning of 1945, when the first elements of schwere Panzerjäger-Abteillung 653 were at the front, to May of the same year, when the last German units surrendered. Note that the national boundaries are not those of the 1940s, where Austria had been incorporated into the Reich and western Czechoslovakia was a German protectorate known as Bohemia and Moravia, but rather those with which present-day readers will be more familiar. Changes to the front line and lost German territory are shown as follows: 1. The front as of 5 January 1945 when it had been stabilised somewhat following the Ardennes Offensive and Operation Nordwind. 2. The Allied advance up to 24 March 1945. 3. The approximate boundaries of the Ruhr Pocket on 5 April 1945, the day after US 1st and 9th Armies completed the encirclement of a large part of Heeresgruppe B including 1.Kompanie and 2.Kompanie, schwere Panzerjäger-Abteilung 512. 4. The front line on 18 April 1945. 5. The furthest extent of the Allied advance as of 8 May 1945 when hostilities in the West officially ended. Note that, in places, this exceeded the boundaries of the demarcation agreed to with the Soviets. Other locations mentioned in the text but not listed by name above include: 6. Sankt Valentin, where the Nibelungenwerk was located. 7. Truppenübungsplatz Döllersheim, where most of the training of the Jagdtiger crews took place. 8. Truppenübungsplatz Sennelager. 9. Zweibrucken. 10. Hagenau. 11. Speyer. 12. Strenberg. 13. Bad Driburg. 14. Offensen. 15. Harste. 16. Beierfelde.

The story of the Jagdtiger and the formations which operated these large tank destroyers is a complex and confusing one as readers will no doubt surmise from the following unit histories. Official documents, such as the various Führungsliste, or allocation lists, are consistently accurate even up to the last weeks of the war and the orders disseminated by Oberkommando des Heeres (OKH), the high command of the army, are also helpful. Some of the most productive sources are the reports prepared for Hitler who was convinced of the vehicle's war-winning potential and followed its progress closely, almost on a daily basis. But it should be remembered that the Jagdtiger operated at a time when Germany was being overrun and the Third Reich was crumbling and many records simply did not survive the conflict. Therefore, much of our knowledge has been gleaned from after-action reports, veterans' accounts and contemporary photographs and the latter, showing as they do mostly disabled or badly damaged vehicles, are representative of the whole. In an effort to simplify the subject, the units discussed here are listed in a rough chronology which combines the allocation of the vehicles and their use in combat. As a matter of convenience I have used the abbreviations s.PzJg.Abt for schwere Panzerjäger-Abteilung and Pz.Ers.u.Aus.Abt to denote Panzer-Ersatz-und Ausbildungs-Abteilung, a training and replacement unit, in this section and throughout this book.

Panzerjäger-Lehr-Abteilung 130. It was planned that this battalion's 3.Kompanie would be the first Jagdtiger unit and it was anticipated that the vehicles would be available by March 1944.

As late as 20 January 1944 a Jagdtiger company was included in the Gliederung published for the Panzer-Lehr-Division, the battalion's parent formation, but it is not mentioned in the documents available for February and the following months, suggesting that the idea had been dropped by then.

The company was to have been allocated fourteen Jagdtigers with two assigned to the headquarters and four with each of the three platoons. This was the same number allotted to the Elefant and Jagdpanther heavy tank destroyer companies but I have been unable to locate a specific KstN (1).

Production delays meant that the company was eventually equipped with Jagdpanzer IV tank destroyers.

s.PzJg.Abt. 653. Originally formed from elements of Sturmgeschütz-Abteilung 197 in March 1943, the battalion was equipped with the Sdkfz 184 Panzerjäger Tiger (P), initially known as the Ferdinand but later referred to as the Elefant. The battalion fought in Russia, taking part in Operation Citadel, and 1.Kompanie later served in Italy but returned to northern Europe in August 1944.

By that time the battalion had been reduced to just twelve vehicles and these were concentrated in 2.Kompanie which would eventually be renamed s.PzJg.Kp 614 and operate as a semi-independent formation (see below). The remainder of the battalion was transferred to Fallingbostel, west of Münster, and under the auspices of Pz.Ers.u.Aus.Abt 500 began training on the first Jagdtigers (2). On 21 September 1944 the crews of 1.Kompanie and 3.Kompanie were moved to Linz in Austria and the men commuted

Notes

1. This subject is discussed further, and the KstN system fully explained, on page 60.
2. A detailed list explaining the allocation of these vehicles begins on page 11.

Left: Photographed at Döllersheim in October 1944, this Jagdtiger, Fgst Nr.305009, was one of the vehicles allocated to s.PzJg.Abt 653 for driver training that stayed with the battalion. Assembled in August 1944 it is fitted with the early-pattern external gun travel lock introduced into production in that month which lacks the horizontal support. Note the Zimmerit anti-magnetic mine paste which covers the hull sides to the level of the highest track hanger, the Porsche suspension and the narrow transport tracks.

Right: Also photographed at Döllersheim in October 1944, this Jagdtiger, Fgst Nr.305010, is painted in an unusual camouflage scheme which may also have been applied to Fgst Nr.305008. The exact significance of the white letter A on the Bosch headlight is uncertain but it has been suggested that it signifies acceptance by one of the Heereswaffenamt Abnahmestelle, or inspection offices, and stamps used by these organisations did feature the letters Wa. A. followed by a number identifying the specific office. This vehicle was also built on the Porsche suspension.

Photographed during the first weeks of 1945, this vehicle of schwere Panzerjäger-Abteilung 653 may be from 3.Kompanie and the presence of Feldwebel Erich Bönike, third from the right, suggests that this may be Jagdtiger number 314. Note the Porsche road wheels and the extensions fitted to the Gg 24/800/300 tracks.

Notes

1. A Jagdtiger training unit had been set up at Munich-Neufreimann as early as June 1944 and these men formed the nucleus of 2.Kompanie but it would be some time before it joined the battalion.
2. Karl-Heinz Münch in *The Combat History of German Heavy Anti-Tank Unit 653 in World War II* suggests that a number of 3.Kompanie vehicles had been in action here since 31 December 1944.
3. It is unclear if this reckoning takes Unteroffizier Jaskiela's Jagdtiger into account.

on a daily basis to the Nibelungenwerk (Ni-Werk) in nearby Sankt Valentin to help with the assembly of the Jagdtiger and familiarise themselves with their new vehicles. But an Allied air raid of 16 October 1944 almost brought production to a halt and the plan to have the two companies fully equipped by the end of December was thwarted. On 1 November 1944 the first and third companies were transferred to Döllersheim, north-east of Linz, and conducted gunnery practice and driver training there until the middle of the month. It was probably at about this time that the formation of a new 2.Kompanie began, under Oberleutnant Robert Wiesenfarth, as it is first mentioned in a report dated 1 December 1944 (1).

On 7 December 1944, 1.Kompanie with seven serviceable Jagdtigers left Döllersheim for Schleiden near the Belgian border to take part in the planned Ardennes Offensive but as rail transport to their starting point was unavailable they arrived too late to be committed to the battle and remained at Blankenheim, 18 kilometres behind the front. On 26 December the Jagdtigers were diverted to the Zweibrücken area with the intention that they should participate in Operation Nordwind. But as the railway lines had been heavily damaged by Allied bombing they were ordered to Reifsdorf near Bonn from where four vehicles were able to be transported to Zweibrücken by 2 January 1945. A further eight Jagdtigers, possibly from 3.Kompanie, which had been sent to the front and were near Wengerohr, almost 80 kilometres south-east of Blankenheim, were ordered to move by road to Boppard, south of Koblenz. These vehicles all suffered mechanical failures and although two were classified as operational in a report from General der Panzertruppen West dated 9 January 1945, some were being recovered and repaired as late as the first week of February.

On 30 December 1944 three Jagdtigers from 1.Kompanie that had been sent from Döllersheim arrived at Zweibrücken and on the same day Hitler personally ordered that they support the attack of 17.SS-Panzergrenadier-Division towards Rimling and Rohrbach, north-east of Aachen, the opening phase of Operation Nordwind. The three Jagdtigers went into action for the first time on 9 January 1945 and achieved mixed results, with one vehicle, numbered 134 commanded by Unteroffizier Fritz Jaskiela, being completely destroyed after it was hit by a Bazooka rocket which penetrated the hull side, detonating the ammunition and killing the entire crew (2).

Late on 10 January 1945 two of the available Jagdtigers were attached to 10.SS-Panzer-Division for an assault between Hatten and Rittershofen and in the ensuing action they proved adept at destroying enemy bunkers.

On 23 January 1945 the first vehicles of 2.Kompanie began arriving at the front and the battalion reported that its full allocation of forty-two Jagdtigers had been fulfilled (3). In the following days, in something of a logistical feat, the battalion was concentrated near Muckenschopf, crossed the Rhine by ferry and occupied a position at Offendorf, just north of Strasbourg. A limited reconnaissance was conducted as far north as Sessenheim, on the Strasbourg to Lauterbourg road, but no contact was made with the enemy. On 29 January, as the battalion was being ferried back across the Rhine, they were attacked by American fighter-bombers which concentrated their fire on one of the newly-arrived Flakpanzer IV anti-aircraft vehicles. Although the Jagdtigers escaped unscathed, Leutnant Arthur Allspach, the commander of the Flak-Zug, and four of his men were killed.

Due to maintenance and supply difficulties operations were curtailed as the Jagdtigers were dispersed among the towns and villages south of Landau until 17 February 1945 when the battalion commander, Major Rudolf Grillenberger, received orders to move to Hagenau in anticipation of an Allied offensive.

The last elements of the battalion arrived in the area north of Hagenau, which was partly in Allied hands, in the first days of March and the Jagdtigers were strung out between Soultz-sous-Forêts, Surbourg, and Morsbronn-les-Bains on the north-eastern edge of what is today the North Vosges Regional Natural Reserve. Major Grillenberger recalled that some of his officers had served in artillery units and suggested that the 12.8cm guns be used in an indirect-fire role. In the late afternoon, under cover of the approaching darkness, the guns began firing on the American positions in Hagenau, directed by a German officer in a forward observation post outside the town, scoring a number of hits in the first exchange. Although the American units returned fire almost Immediately, the gunners were safe inside their massive vehicles. For the first two weeks of March 1945 the battalion held the positions north of Hagenau which took on a semi-permanent look with sturdy bunkers constructed around the vehicle positions. But American penetrations towards Frankfurt, far to the north, meant that the German units around Hagenau would need to retreat to avoid being cut off and s.PzJag.Abt 653 was ordered to cover the withdrawal.

Major Grillenberger received the order on 14 March and by that evening the battalion was involved in heavy fighting. The battle continued into the next day when 3.Kompanie was forced to break off a counterattack against American positions at Hollendorf after losing five Jagdtigers to Allied fighter-bombers.

On 16 March 1945 the Jagdtigers of Oberleutnant Haberland's 1.Kompanie were able to push the Americans back in the area west of Surbourg but not without loss and the battalion began falling back towards the relative safety of the Westwall defences. Only thirteen of the battalion's Jagdtigers were able to make the journey on their own tracks and the remaining twenty-one were moved by rail or in desperate towing operations (1).

The mobile elements of the battalion that reached the German frontier were ordered to delay the advancing Allies and during the evening of 21 March 1945 a number of Jagdtigers from 2.Kompanie, led by Leutnant Feineisen, took up positions in the small town of Böhl (2), between the road to Ludwigshafen and the Rhine, and waited for the Americans.

Notes

1. Major Grillenberger had earlier complained that towing a single Jadgtiger required the efforts of a Bergepanther recovery tank and two Sdkfz 9 halftracks.

2. Today the town is known as Böhl-Iggelheim but Böhl and Iggelheim were two separate entities in 1945.

Photographed in Neustadt an der Weinstrasse on Landauer Strasse near the corner of Speyerdorfer Strasse on 21 March 1945, Jagdtiger 331 from 3.Kompanie of schwere Panzerjäger-Abteilung 653 was commanded by Leutnant Kaspar Göggerle and had been engaged in a fierce firefight in the open country outside the town. Athough it was hit at least six times Göggerle still managed to withdraw towards the cover of the buildings seen here but ran out of ammunition and his vehicle was abandoned together with Jagdtiger 323.

The enemy units appeared at around 8.00am the following morning and in the ensuing firefight nine tanks and two halftracks were destroyed. Another two halftracks were captured intact and thirteen Americans taken prisoner. Leutnant Feineisen's Jagdtiger broke down but was towed away under fire, repaired and returned to the battle. On the same day, less than 3 kilometres to the west, a Kampfgruppe from 3.Kompanie led by Leutnant Kaspar Göggerle engaged an American armoured column near Neustadt an der Weinstrasse and later claimed to have knocked out twenty-five enemy tanks. But two Jagdtigers, number 331 and number 323, were abandoned after they broke down.

Heavy fighting continued in Böhl where Leutnant Feineisen was wounded twice and although his vehicle survived, the other two Jagdtigers of his Kampfgruppe were lost. Just to the north of Feineisen's position the Jagdtigers of 1.Kompanie were located in the town of Schifferstadt protecting the Rhine crossing at Speyer and held back a strong American attack throughout the day, falling back across the river as night approached with the Jagdtigers of 2.Kompanie. A number of damaged vehicles that Major Grillenberger had been forced to leave at Bellheim further to the south were transported by rail across the Rhine at Germersheim under cover of darkness. After the last vehicles crossed the river at Speyer the bridge was destroyed.

On 22 March 1945, the day the Jagdtigers of s.PzJg.Abt 653 were fighting to hold the approaches to Speyer, almost 100 kilometres to the north American units were able to create a strong bridgehead at Oppenheim and used this to push towards Heidelberg and Mannheim, bringing the latter under artillery fire by 28 March. That afternoon the eight operational Jagdtigers were ordered to the area north of Heidelberg to support elements of 559.Volksgrenadier-Division which were holding a defensive line on the northern bank of the Neckar Canal. A report prepared by the headquarters of LXXX.Armeekorps mentions that a successful counterattack was made in this area on Thursday, 29 March 1945 involving an unspecified number of Jagdtigers (1).

But by this time the enemy had crossed the canal in several places and when three Jagdtigers of 1.Kompanie approached Schwetzingen, a few kilometres to the west of Heidelberg, at around noon on Friday they did not realise that the Americans were already in possession of the town. The vehicle of Oberfeldwebel Ferdinand Kinnberger, number 131, was hit several times at close range and began burning immediately while the company commander's Jadgtiger, number 115, was bogged down on the northern edge of the town and had to be abandoned. The third Jagdtiger, commanded by Oberfeldwebel Gustav Koss, managed to escape but only as far as Eppelheim on the Heidelberg to Mannheim road where it was hit in the track and could not be repaired.

In addition to the combat losses, one Jagdtiger was totally destroyed on 31 March 1945 during routine maintenance of the engine's air pre-filter (2).

On 3 April 1945, as part of the general withdrawal, Major Grillenberger moved the battalion headquarters to Cleeborn, north of Stuttgart. Covering the battalion, two Jadgtigers of 1.Kompanie moved from

Notes

1. This report is rather vague, merely mentioning 'our Jagdtigers'.
2. Ironically this accident accounted for more fatalities than the action in front of Schwetzingen.

Commanded by Feldwebel Heinz Telgmann, Jagdtiger 332 of 3.Kompanie was abandoned after it was hit by a number of rockets fired by Allied fighter-bombers near Morsbronn-les-Bains in mid-April 1945 when schwere Panzerjäger-Abteilung 653 was defending the approach to the Speyer bridge on the Rhine. Feldwebel Erich Bönike's Jagdtiger 314 and Jagdtiger 301 commanded by Unteroffizier Hans Appel were lost in the same action.

Jagdtiger Fgst Nr.305083 photographed at the Red Army's proving ground at Kubinka near Moscow shortly after the war. This vehicle is one of four that were picked up from the Nibelungenwerk in late April 1945 by crews from schwere Panzerjäger-Abteilung 653 as mentioned in the text. It is unlikely that the elaborate camouflage scheme, complete with accurately-sized Balkenkreuz markings on the superstructure sides and rear, would have been applied by the Russians and we can be reasonably sure that the appearance of this vehicle has altered little since it left the factory. Note the late-pattern ribbed front fenders which were introduced into the Tiger II production run in March 1945 and may have been incorporated into the Jagdtiger programme at about the same time, although this photograph provides the only known example.

Eppingen towards the village of Kirchardt where the vehicle commanded by Feldwebel Reinhold Schlabs promptly broke down. It was towed by the other Jagdtiger, with Leutnant Knippenberg on board, back to Eppingen where the second vehicle experienced the same mechanical difficulties. The crews waited there for the maintenance unit to arrive but as night fell they clearly heard the sound of approaching enemy tanks and destroyed both vehicles.

By this time the battalion was not operating as a cohesive unit and one or two vehicles were allocated to defensive points as they were needed. The fighting around the village of Klinenberg, outside Heilbron on the Neckar River, on 5 April 1945 was particularly heavy and one Jagdtiger was lost to enemy action while a second was destroyed by its own crew.

Two days later a Kampfgruppe from the battalion managed to stabilise the front near Crailsheim, claiming the destruction of many enemy tanks, but by Friday, 13 April 1945 the survivors had withdrawn to Nördlingen, north of Augsburg. On the following Sunday Major Grillenberger left the battalion to be replaced by Major Rolf Fromme who had previously led schwere Panzer-Abteilung 503 (1).

On 20 April any crews without vehicles were ordered to report to the Linz arsenal to pick up new Jagdtigers as they were assembled at Ni-Werk in nearby Sankt Valentin. An OKH order of 17 April 1945 mentions that three Jagdtigers were held at the Linz arsenal for s.PzJg.Abt 653 but the Führungsliste clearly shows four Jagdtigers allocated to the battalion in the same month, the last recorded shipment of the war. Feldwebel Rheinhold Schlabs, who commanded one of the Jagdtigers, was adamant that four were loaded onto railway cars and transported towards the front arriving at Sankt Leonard am Forst on Tuesday, 1 May 1945. Feldwebel Schlabs recounted how one Jagdtiger broke down and was left behind but the rest continued to Strengberg where they eventually surrendered. One of these vehicles is today on display at the Kubinka Tank Museum near Moscow.

On 2 May 1945 an OKH order advised that the battalion would be subordinated to Heeresgruppe Süd and would move immediately to Linz via Salzburg and it is indicative of the general situation that companies are no longer referenced but rather individual vehicles are mentioned. The fighting element of the battalion at that time consisted of a Kampfgruppe led by Oberleutnant Franz Kretschmer made up of Jagdtigers 312 and 324 (2).

The two 3.Kompanie vehicles, together with the men of the battalion's supply and maintenance units, reached Liezen in central Austria on 7 May 1945 where they surrendered to units of the US Army. Earlier, on 4 May, troops under the direction of Leutnant Hans Knippenberg, who had led the battalion's 1.Kompanie, destroyed the last eight Jagdtigers at the Nibelungenwerke.

A detailed table of organisation, based on a surviving Gliederung for s.PzJg.Abt 653, is shown on pages 60 and 61 and the battalion's actual organisation at the beginning of March 1945, when some of the heaviest fighting occurred, is presented on page 62.

Notes

1. Grillenberger had been removed from his command, officially at least, on 25 March 1945 and posted to the Führerreserve. Hitler personally ordered his demotion to Leutnant and he ended the war as a platoon commander with schwere Panzer-Abteilung 507.
2. As stated, these vehicles are specifically mentioned in the 2 May 1945 order but veterans' accounts suggest that at least one Jadgtiger of 1.Kompanie and another from 2.Kompanie were still operational at this time but separated from their parent formation.

s.PzJg.Kp 614. As mentioned earlier, s.Pz.Abt 653 was originally equipped with the Sdkfz 184 Panzerjäger Tiger (P), better known as the Elefant. By August 1944 the battalion had been reduced to just twelve tank destroyers and these, with two reinforcements, were concentrated in 2.Kompanie under Oberleutnant Werner Salamon and after refitting at Krakow were sent to the front and subordinated to 17.Armee.

By late November, Hauptmann Bernhard Konnak had arrived to take command and in the following December the company was renamed schwere Panzerjäger-Kompanie (s.PzJg.Kp) 614. Fighting in the defence of Kielce, between Krakow and Lublin, the company lost eight Elefants and the remaining four were badly damaged. Following a series of conflicting directives regarding the company's fate, an OKH order dated 22 February 1945 stated that s.PzJg.Kp 614 was to be rebuilt at Stahnsdorf in Germany in accordance with KstN and KAN 1148c with amendments as outlined in AHM 44 number 488 (1).

The 22 February order describes KstN 1148c as 's.PzJg.Kp (zu 10 Elefanten bezw. Jagdtiger, Porsche Laufwerk)', that is, 'a heavy anti-tank company with ten Elefants or to be more precise Jagdtigers with Porsche suspension'. KstN 1148c of 31 March 1943 was the establishment used to build the Ferdinand/Elefant companies that fought in Russia and Italy and its inclusion here is interesting for two reasons. Firstly, it suggests that specific KstN may not have been created for the Jagdtiger companies but rather that amended versions of the 1943 organisation were used. And secondly, the allocation of ten Jagdtigers fitted with the Porsche suspension could only have been acheived in February 1945 if either all the serviceable vehicles dispatched up to that time were recalled or if production of the type was resumed.

s.PzJg.Abt 512. Formed in early February 1945 from elements of schwere Panzer-Abteilung 424, a Tiger unit which had lost most of its tanks in a single action near the town of Lisow in eastern Poland (2), the battalion was commanded by Major Walter Scherf.

The first and second companies began training at Döllersheim and received the first five Jagdtigers on Monday, 12 February with another six shipped on the following Wednesday. But within days an OKH order directed that the steering gear of all Jagdtigers awaiting delivery or on hand with units was to be replaced. Subsequently five vehicles, possibly the initial allocation, went to the battalion's workshop at Döllersheim while work on a further six was carried out at the nearby Linz arsenal. It is unclear if all eleven vehicles that had been in repair were returned to the battalion and the allocation entry for the initial five were in fact struck off on the Führungsliste.

On 3 March 1945, 2.Kompanie under Oberleutnant Otto Carius was transferred to Sennelager, outside Paderborn in western Germany, and ten days later they had been joined by 1.Kompanie, commanded by Oberleutnant Albert Ernst.

This Jagdtiger, Fgst Nr.305058, was one of the vehicles recalled in February 1945 due to problems with the steering gear. It was later assigned to 2.Kompanie, schwere Panzerjäger-Abteilung 512 and numbered X7. It was disabled, possibly by friendly fire, and abandoned near the corner of Bahnhofstrasse and Oranienstrasse on the western edge of Netphen, north-east of Siegen.

Notes

1. AHM, or Allgemeine Heeresmitteilungen, were army news bulletins which were published on a regular basis and contained information on organisational and name changes, administrative directions and general orders. They were identified by the year, 44 in this case, and numbered consecutively and although we have no date for this example the number 488 would suggest that it was issued in mid-August 1944.
2. The history of this unit is covered in *Tank Craft 31: Tiger I and Tiger II Tanks German Army and Waffen-SS The Last Battles in the East, 1945.*

The last three Jagdtigers of 1.Kompanie, schwere Panzerjäger-Abteilung 512 photographed on 16 April 1945 in the Schillerplatz at Iserlohn where they surrendered to units of the US Army. The vehicle closest to the camera is thought to be number X8 commanded by Leutnant Sepp Tarlach, with Leutnant Heinz Rondorfs's X2 in the centre. Note the kill rings on the barrel of the latter. The next Jagdtiger in line is easily identifiable as X1, the vehicle of the company commander Hauptmann Ernst, by the Sternantenne command radio aerial just visible behind the crewmen.

A strength report prepared for the Generalinspekteur der Panzertruppen (Gen.Insp.d.Pz.Tr.) dated 5 March 1945 stated that ten Jagdtigers had been sent from Linz to Sennelager by that date and that these were intended for 1.Kompanie. A further three vehicles were at the Linz arsenal while another five were held at the Sankt Valentin factory awaiting tracks and two more were expected to be fully assembled by 8 March. All ten were earmarked for 2.Kompanie.

In any case Major Scherf reported that when the last elements of 1.Kompanie arrived in Sennelager on 13 March 1945, there were now twenty Jagdtigers on hand. On the same day the formation of the battalion's 3.Kompanie began at Döllersheim and the Gen.Insp.d.Pz.Tr. report had earlier mentioned that nine Jagdtigers intended for this company were expected to be built by 10 March.

It was planned that the Jagdtigers would take part in the assault on the American bridgehead at Remagen and although the battalion's light vehicles left Sennelager as early as 14 March 1945 it was another six days before the eighteen serviceable Jagdtigers were loaded onto railway cars and these did not arrive at Olpe, almost 60 kilometres from the front, until 26 March. Once deployed, the heavy tank destroyers were plagued by mechanical faults and of the seven Jagtigers lost in rearguard actions, four were due to breakdowns. But in a later skirmish around Herborn, between Wetzlar and Siegen, when the German units were fighting from fixed positions, the 12.8cm guns of the Jagdtigers were able to hit enemy tanks at ranges of up to 3,000 metres, claiming the destruction of thirty vehicles.

On 1 April 1945 the lead elements of two Allied Army Groups which had been pushing into Germany met at Lippstadt and within days 370,000 German soldiers, including the two Jagdtiger companies, were encircled in what became known as the Ruhr Pocket. In the first weeks of April, the remnants of 1.Kompanie formed part of a Kampfgruppe, led by Oberleutnant Ernst, made up of the remaining Jagdtigers with four Sturmgeschütz III assault guns, four Pzkpfw IV tanks and an anti-aircraft platoon (1) and it was probably at about this time that Ernst was promoted to Hauptmann. Oberleutnant Carius' 2.Kompanie was moved to Unna, east of Dortmund, and took part in the defence of the city.

An OKH order of 31 March 1945 released 3.Kompanie, which had moved from Döllersheim to Sennelager with five Jagdtigers, for deployment by OB West. The same order stated that two more vehicles were in transit from Sankt Valentin and that the repairs on a third were expected to be completed that day. By this time units of the US Army were just a few kilometres away and the company fought briefly around Paderborn, losing one Jagdtiger there on 2 April 1945, and it is possible that a single Jagdtiger was left at Sennelager. The remaining vehicles withdrew towards the east and on 8 April the company commander, Oberleutnant Werner Schrader (2), was killed while

Notes

1. Some accounts state that the assault guns were Sturmgeschütz IV but no allocations are mentioned in the relevant Führungsliste for either model nor for the Pzkpfw IV tanks. This would seem to confirm that the assault guns and tanks were not in fact an integral part of the battalion, as is often suggested, but rather seconded from units in the area. The anti-aircraft platoon almost certainly came from 116.Panzer-Division.
2. Sometimes spelt Schräder.

conducting a reconnaissance near Volkmarshausen on the Weser River. On the following day one Jagdtiger was abandoned near the village of Offensen and in a counterattack at Harste in support of the Tigers of schwere Panzer-Abteilung 507 another vehicle was totally burnt out.

The last two Jagdtigers of 3.Kompanie fought with SS-Brigade Westfalen in the battles between Beierfelde and Sankt Andreasberg in the Harz Mountain region and were lost on 11 April 1945.

On 15 April 1945 Oberleutnant Carius ordered the destruction of the last six Jagdtigers of his company and the following day Hauptmann Ernst, with three surviving Jagtigers, surrendered at Iserlohn, south-east of Dortmund.

A table of organisation showing the battalion's actual strength in early 1945 is presented on page 63.

Panzer-Kompanie Kummersdorf. This ad-hoc formation was created on 31 March 1945 and was to be built around the assets of Versuchs-Abteilung Waffenamt A, a unit based at the weapons trial centre at Kummersdorf, south of Berlin. The company was made up of five mixed platoons and the three Panzer-Züge, described as 'bedingt beweglich' or conditionally mobile, contained a Jagdtiger, a Tiger II, four Panthers, two Pzkpfw IV, a single Pzkpfw III and two US Army Sherman tanks. On 19 April the company was moving south towards Luckau, about 60 kilometres south of Berlin, but two days later it was ordered to join Kampfgruppe Möws which at that time was at Zossen, just outside the capital. Although I cannot offer any information as to origin of the Jagdtiger, or its fate, it is most certainly not Fgst Nr.305004 as claimed in some accounts. The latter was captured intact near Sennelager on, or just before, 6 April 1945, a distance of approximately 460 kilometres from Kummersdorf (1).

Schwere SS-Panzer-Abteilung 501. On 2 May 1945 about forty men of the battalion were sent to the Nibelungenwerk at Sankt Valentin with orders to make operational six Jagdtigers which were at the factory. Two vehicles returned to the headquarters of 1.SS-Panzer-Division which was at that time at Scheibbs, about 45 kilometres to the east of Sankt Valentin, and they were immediately committed to an attack towards Waidhofen an der Ybbs. On 8 May 1945 one of the Jagdtigers was abandoned after losing a track while attempting to cross a bridge and on the following day the remaining vehicle was disabled and used as a roadblock on the southern edge of the town.

The post-war memoirs of the battalion adjutant, Klaus-Peter Schwab, have been used by some authors to suggest that the 8.8cm armed version of the Jagdtiger saw service with this battalion but Schwab explains in his account that the vehicles he was promised on 4 May 1945 were destroyed at the factory (2).

Notes

1. Karl-Heinz Münch in *The Combat History of German Heavy Anti-Tank Unit 653 in World War II* states that the Jagdtiger of Panzer-Kompanie Kummersdorf had some connection to s.PzJg.Abt 653 but does not elaborate.

2. Schwab claimed that he was ordered to take control of ten to twelve Jagdtigers from an 'Austrian factory' and while there are several obvious inaccuracies in his description of events, such as the reference to Speyer as the location of the factory or the '12,2cm Kanone', these could be simple mistakes. He does make a point of saying that the vehicles were armed with the 8.8cm gun in place of the larger weapon and it is difficult to explain how he could have known this was a possibility if the information was not included in his orders.

Said to have been photographed near Sankt Andreasberg in the Harz Mountain region, this Jagdtiger is almost certainly one of the last vehicles of 3.Kompanie, schwere Panzerjäger-Abteilung 512. The Pilze socket just above the Balkenkreuz, identifying this as a January 1945 production vehicle, would also point to that unit. The camouflage scheme is unusual as is the number on the superstructure side which appears to be 35 or 55. Another Jagdtiger abandoned in the same area had a white number on the hull side which may have been 11 or I1 but it is very badly worn.

THE JAGDTIGERS, 1944-1945

As these vehicles were produced in relatively small numbers I have endeavoured to show some of the production and operational history of each Jagdtiger. All armoured vehicles were identified by a Fahrgestellnummer (Fgst Nr.), or chassis number, and blocks of these numbers were allocated by the Heereswaffenamt when a contract was awarded and those of the Jagdtiger programme began with 305001. The Jagdtigers were all assembled at the Nibelungenwerke, often referred to as Ni-Werk, built near the village of Sankt Valentin close to the Linz arsenal and about 25 kilometres from Döllersheim where much of the training took place. Vehicles requiring extensive repair were returned to Ni-Werk and it is probable that a number were there when the factory was overrun by the Red Army. Although the Führungsliste, or allocation lists, are accurate to at least April 1945, the same cannot be said of the production statistics and much information was lost in the war. Although photographs show that hulls were constructed for at least ninety-two vehicles it is likely that not more than eighty-five Jagtigers armed with the 12.8cm were completed. As earlier, I have used the abbreviations s.PzJg.Abt for schwere Panzerjäger-Abteilung and Pz.Ers.u.Aus.Abt to denote Panzer-Ersatz- und Ausbildungs-Abteilung, a training and replacement unit.

Fgst Nr.	Assembled	Suspension	Remarks
305001	February 1944	Porsche	Sent to Waffenprüfamt VI at Haustenbeck, the army's weapons testing authority, it was damaged and replaced, probably at some time after July 1944, by Fgst Nr.305004. During repairs a three-colour camouflage pattern was applied and tool racks and twelve track hangers were eventually added to the hull sides. It was later assigned to Pz.Ers.u.Aus.Abt 17, a unit specifically raised to train Jagdtiger crews, and then allocated to s.PzJg.Abt 653 and may have been one of the vehicles that operated with a Kampfgruppe of 1.SS-Panzer-Division in the last days of the war.
305002	February 1944	Henschel	Sent to the weapons testing centre at Kummersdorf and later assigned to Pz.Ers.u.Aus.Abt 17 and then to s.PzJg.Abt 653. It was eventually abandoned near Weissenbach in early May 1945.
305003	June 1944	Porsche	This vehicle was fitted with covers for the exhaust mufflers. The hull was coated with Zimmerit anti-magnetic mine paste on all vertical surfaces including the muffler covers. The Zimmerit was applied on the hull sides to the level of the lowest track hanger. Retained at Ni-Werk to test the feasibility of using Sdkfz 184 Elefant tracks, it was later sent to the Panzerjäger-Schule at Mielau. It is tempting to speculate that this vehicle may be the Jagdtiger reported as a loss by the headquarters of 11.Armee, fighting on the Oder Front, in mid-February 1945. Delivered on the last day of June and categorised as a July production model in some accounts.
305004	July 1944	Porsche	This vehicle was one of just two that were fitted with covers for the exhaust mufflers. Zimmerit was applied in the same manner as described for Fgst Nr.305003. It replaced Fgst Nr.305001 with Waffenprüfamt VI and is often said, incorrectly, to have served with Panzergruppe Kummersdorf in the last weeks of the war. It is today on display at the RAC Tank Museum, Bovington.
305005	July 1944	Porsche	Zimmerit was applied in the same manner as described for Fgst Nr.305003 but this vehicle lacked the muffler covers. It was sent to Pz.Ers.u.Aus.Abt 500 at Fallingbostel where its armour was found to be defective. It was retained by the army for homeland service.
305006	August 1944	Porsche	Zimmerit was applied in the same manner as described for Fgst Nr.305003 but this vehicle lacked the muffler covers. This vehicle was initially sent to Pz.Ers.u.Aus.Abt 500 and later allocated to s.PzJg.Abt 653 for driver training and remained with the battalion.
305007	August 1944	Porsche	Zimmerit was applied in the same manner as described for Fgst Nr.305003 but this vehicle lacked the muffler covers. Allocated to s.PzJg.Abt 653 for driver training from Pz.Ers.u.Aus.Abt 500 and remained with the battalion.
305008	August 1944	Porsche	Allocated to s.PzJg.Abt 653 for driver training from Pz.Ers.u.Aus.Abt 500 and remained with the battalion. Zimmerit was applied to the vertical surfaces and may have been extended upwards on the hull sides to the level of the top track hangers. This would provide a possible identification as Jagdtiger number 102, the vehicle shown in the photograph on page 13.

..........table continues on next page

The only known photograph of Fgst Nr.305002, the first Jagdtiger fitted with the Henschel suspension. Note that the modifications incorporated into series production, such as the track hangers, tool racks and external gun travel lock, are absent here.

Fgst Nr.	Assembled	Suspension	Remarks
305009	August 1944	Porsche	The Zimmerit was applied on the hull sides to the level of the highest track hanger. Allocated to the Ersatzheer and then s.PzJg.Abt 653 for driver training, it remained with the battalion. Assigned to 2.Kompanie it was abandoned outside Nordheim on the road to Lauffen am Neckar in early 1945 together with Jagdtiger number 214, Fgst Nr.305037.
305010	August 1944	Porsche	This was the last vehicle coated in Zimmerit. It was applied in the same manner as described for Fgst Nr.305009. Allocated to the Ersatzheer and then s.PzJg.Abt 653 for driver training, it remained with the battalion and was assigned to 3.Kompanie where it was numbered as 301. On 16 March 1945, near Höllendorf, the vehicle was caught in an artillery barrage and after receiving a direct hit to the radio operator's hatch caught fire and was eventually burnt out.
305011	September 1944	Porsche	Allocated to the Ersatzheer and then s.PzJg.Abt 653, this vehicle was assigned to 1.Kompanie and probably numbered as 115. It was disabled near Schwetingen in late March 1945.
305012	September 1944	Porsche	This was the last production vehicle built with the Porsche suspension (see 305049). It was allocated to the Ersatzheer and then s.PzJg.Abt 653 and assigned to 3.Kompanie where it was numbered as 314. It was destroyed by its own crew in mid-March 1945.
305013	September 1944	Henschel	Allocation unclear, may have been damaged in the October bombing.
305014	September 1944	Henschel	Allocated to s.PzJg.Abt 653 in October 1944.
305015	September 1944	Henschel	Allocated to s.PzJg.Abt 653 in October 1944.
305016	September 1944	Henschel	Allocated to s.PzJg.Abt 653 in October 1944.
305017	September 1944	Henschel	Allocated to s.PzJg.Abt 653 in October 1944.
305018	September 1944	Henschel	Allocated to s.PzJg.Abt 653 in October 1944.
305019	September 1944	Henschel	Allocated to s.PzJg.Abt 653 in October 1944.
305020	September 1944	Henschel	Allocated to s.PzJg.Abt 653 in November 1944, this vehicle was assigned to 3.Kompanie and numbered as 331. It broke down near Neustadt on 22 March 1945 and was abandoned. It survived the war and is on display at the US Army Ordnance Museum, Aberdeen.
305021	October 1944	Henschel	Allocated to s.PzJg.Abt 653 in November 1944.
305022	October 1944	Henschel	This vehicle was allocated to s.PzJg.Abt 653 in November and assigned to 3.Kompanie as Jagdtiger number 323. It was abandoned near Neustadt at the same time as Fgst Nr.305020 mentioned above.
305023	October 1944	Henschel	Assembled in October 1944, this vehicle was allocated to s.PzJg.Abt 653 in November and assigned to 1.Kompanie as Jagdtiger number 113. It was destroyed by its own crew on 4 May 1945 near Rosenheim.
305024	October 1944	Henschel	This vehicle was allocated to s.PzJg.Abt 653 in November 1944 and assigned to 1.Kompanie and numbered as 134. It was destroyed near Rimling on 9 January 1945 with the loss of the entire crew.
305025	October 1944	Henschel	This vehicle was allocated to s.PzJg.Abt 653 in November 1944.
305026	November 1944	Henschel	This vehicle was allocated to s.PzJg.Abt 653 in November 1944.
305027	November 1944	Henschel	This vehicle was allocated to s.PzJg.Abt 653 in November 1944.

..........*table continues on next page*

Fgst Nr.305001, the first series production Jagdtiger built with the Porsche suspension. Originally assigned to Waffenprüfamt VI in February 1944, it was damaged and replaced at some time after July by Fgst Nr.305004. In its original configuration this vehicle was not fitted with the spare tracks and their hangers, the tools, towing cable, external gun travel lock or later 9-tooth drive sprocket. The three-colour camouflage and Balkenkreuz, the latter just visible behind the spare track lengths, are also later additions.

Fgst Nr.	Assembled	Suspension	Remarks
305028	November 1944	Henschel	This vehicle was allocated to s.PzJg.Abt 653 in November 1944.
305029	November 1944	Henschel	This vehicle was allocated to s.PzJg.Abt 653 in November 1944.
305030	November 1944	Henschel	This vehicle was allocated to s.PzJg.Abt 653 in November 1944.
305031	November 1944	Henschel	Allocated to s.PzJg.Abt 653 in December 1944, it was probably the 1.Kompanie vehicle numbered as 131 which was disabled and abandoned in Schwetingen on 30 March 1945.
305032	December 1944	Henschel	Allocated to s.PzJg.Abt 653 in December 1944, this Jagdtiger was assigned to 3.Kompanie. It was the first production model to feature the extra track hangers on the superstructure sides.
305033	December 1944	Henschel	Allocated to s.PzJg.Abt 653 in December 1944.
305034	December 1944	Henschel	Allocated to s.PzJg.Abt 653 in January 1945.
305035	December 1944	Henschel	Allocated to s.PzJg.Abt 653 in January 1945.
305036	December 1944	Henschel	Allocated to s.PzJg.Abt 653 in January 1945.
305037	December 1944	Henschel	Allocated to s.PzJg.Abt 653 in January 1945, it was assigned to 2.Kompanie and numbered as 214. It was was disabled and abandoned outside Nordheim together with Jagdtiger Fgst Nr.305009.
305038	December 1944	Henschel	Allocated to s.PzJg.Abt 653 in January 1945.
305039	December 1944	Henschel	Allocated to s.PzJg.Abt 653 in January 1945.
305040	December 1944	Henschel	Allocated to s.PzJg.Abt 653 in January 1945.
305041	December 1944	Henschel	Allocated to s.PzJg.Abt 653 in January 1945.
305042	December 1944	Henschel	Allocated to s.PzJg.Abt 653 in January 1945.
305043	December 1944	Henschel	Allocated to s.PzJg.Abt 653 in January 1945.
305044	December 1944	Henschel	Allocated to s.PzJg.Abt 653 in January 1945.
305045	December 1944	Henschel	Allocated to s.PzJg.Abt 653 in January 1945.
305046	December 1944	Henschel	Allocated to s.PzJg.Abt 653 in January 1945.
305047	December 1944	Henschel	Allocated to s.PzJg.Abt 653 in January 1945.
305048	December 1944	Henschel	Allocated to s.PzJg.Abt 653 in January 1945.
305049	December 1944	Porsche	Shipped to Pz.Ers.u.Aus.Abt 500 at Mielau in December 1944. May have been one of the vehicles damaged in the October 1944 bombing and repaired with Porsche suspension units.
305050	December 1944	Henschel	Allocated to s.PzJg.Abt 512 in February 1945.
305051	December 1944	Henschel	Allocated to s.PzJg.Abt 512 in February 1945.
305052	January 1945	Henschel	Allocated to s.PzJg.Abt 512 in February 1945.
305053	January 1945	Henschel	Allocated to s.PzJg.Abt 512 in February 1945.

..........table continues on next page

Photographed at some time after the end of the war, this Jagdtiger was disabled and abandoned on 17 March 1945 near Rittershofen, north of Strasbourg. Assigned to 1.Kompanie, schwere Panzerjäger-Abteilung 653, this is one of the August 1944 production vehicles and may be Fgst Nr.305008 as explained on page 11. Note the application of the Zimmerit anti-magnetic mine paste and the elaborate camouflage scheme.

Fgst Nr.	Assembled	Suspension	Remarks
305054	January 1945	Henschel	Assembled in January 1945, this vehicle was allocated to s.PzJg.Abt 512 in February 1945. Photographs of the partly-assembled vehicle suggest that it may have been a Befehlspanzer command vehicle.
305055	January 1945	Henschel	Allocated to s.PzJg.Abt 512 in February 1945.
305056	January 1945	Henschel	Allocated to s.PzJg.Abt 512 in February 1945.
305057	January 1945	Henschel	Allocated to s.PzJg.Abt 512 in February 1945. This was one of the vehicles affected by the decision to replace the steering gear, as mentioned below, but it was later assigned to 1.Kompanie and numbered as X5.
305058	January 1945	Henschel	Allocated to s.PzJg.Abt 512 in February 1945. This was one of the vehicles affected by the decision to change the steering gear. It was later assigned to 1.Kompanie and numbered as X7.
305059	January 1945	Henschel	Allocated to s.PzJg.Abt 512 in February 1945.
305060	January 1945	Henschel	Allocated to s.PzJg.Abt 512 in February 1945.

An order dated 16 February 1945 directed that the steering gear of vehicles awaiting delivery or on hand with units was to be replaced. In the case of s.PzJg.Abt 512, five Jagdtigers were repaired by the battalion's workshop at Döllersheim while work on a further six was carried out at the Linz arsenal. These vehicles were not returned to the manufacturer as some sources suggest. It is known that this affected Fgst Nr.305055, 305056, 305057, 305058, 305059 and 305060.

Fgst Nr.	Assembled	Suspension	Remarks
305061	January 1945	Henschel	Allocated to s.PzJg.Abt 512 in March 1945.
305062	February 1945	Henschel	Allocated to s.PzJg.Abt 512 in March 1945.
305063	February 1945	Henschel	Allocated to s.PzJg.Abt 512 in March 1945.
305064	February 1945	Henschel	Allocated to s.PzJg.Abt 512 in March 1945.
305065	February 1945	Henschel	Allocated to s.PzJg.Abt 512 in March 1945.
305066	February 1945	Henschel	Allocated to s.PzJg.Abt 512 in March 1945.
305067	February 1945	Henschel	Allocated to s.PzJg.Abt 512 in March 1945.
305068	February 1945	Henschel	Allocated to s.PzJg.Abt 512 in March 1945. Probably assigned to 2.Kompanie, this may have been the Jagdtiger commanded by Leutnant Hans-Jürgen Haas.
305069	February 1945	Henschel	Allocated to s.PzJg.Abt 512 in March 1945.
305070	February 1945	Henschel	Allocated to s.PzJg.Abt 512 in March 1945.
305071	February 1945	Henschel	Allocated to s.PzJg.Abt 512 in March 1945.
305072	February 1945	Henschel	Allocated to s.PzJg.Abt 512 in March 1945.
305073	February 1945	Henschel	Allocated to s.PzJg.Abt 512 in March 1945.
305074	February 1945	Henschel	Allocated to s.PzJg.Abt 512 in March 1945. It was assigned to 2.Kompanie and knocked out in early April 1945 near Veriehausen, east of Paderborn.
305075	March 1945	Henschel	Allocated to s.PzJg.Abt 512 in March 1945.
305076	March 1945	Henschel	Allocated to s.PzJg.Abt 512 in March 1945.
305077	March 1945	Henschel	Allocated to s.PzJg.Abt 512 in March 1945.

...........table continues on next page

Photographed at Haustenbeck, where it had served with Waffenprüfamt VI, Jagdtiger Fgst Nr.305004 awaits transport to Bremerhaven and then to the UK. The cover for the exhaust muffler is just visible here.

Fgst Nr.	Assembled	Suspension	Remarks
305078	March 1945	Henschel	This vehicle, and Fgst Nr.305079, may have been the Jagdtigers seconded to s.SS.Pz.Abt 101 on 4 May 1945. This is explained further in the section dealing with unit histories. Interestingly, it is also claimed to have been one of the hulls fitted with the 8.8cm Pak 43 L/71 gun as were Fgst Nr. 305079, 305080 and 305081 (see below).
305079	March 1945	Henschel	See Fgst Nr.305078 above.
305080	March 1945	Henschel	See Fgst Nr.305078 above. Probably one of the vehicles destroyed at the Nibelungenwerke to prevent them falling into the hands of the Soviets.
305081	March 1945	Henschel	See Fgst Nr.305078 above.

The use of the 8.8cm gun in the Jagdtiger programme is the subject of extensive debate. In February 1945 Hitler ordered that all possible measures were to be taken to increase Jagdtiger production, including the use of weapons other than the 12.8cm gun, specifically the 8.8cm L/71. A report compiled for the Inspekteur der Panzertruppen and dated 29 April 1945 clearly states that deliveries made since 4 April had consisted of four 'Jagdtiger 12.8' to s.Pz.Jg.Abt 653 and approximately four 'Jagdtiger 8.8'. The report does not elaborate on the final destination of the latter. It goes on to mention that a further seventeen 'Jagdtiger 8.8' were expected to be available in May. In Jagdtiger, The Most Powerful Armoured Fighting Vehicle of World War II, *author and researcher Andrew Devey states that Fgst Nr.305078 to 305081 were built with 8.8cm weapon but that none were issued to units as there were no sights for the guns. He goes on to say that Fgst Nr.305085 to 305088 were to be fitted with 8.8cm weapons but they were probably not completed before 4 May 1945 when the factory was abandoned. The story of Fgst Nr.305081, said to have been fitted with an 8.8cm gun and discovered in a forest in Poland in 1996, is a hoax.*

305082	April 1945	Henschel	Allocated to s.PzJg.Abt 653 in April 1945.
305083	April 1945	Henschel	Allocated to s.PzJg.Abt 653 in April 1945. Captured by the Soviets at Strenberg, this vehicle is today on display at the Kubinka Tank Museum near Moscow.
305084	April 1945	Henschel	Allocated to s.PzJg.Abt 653 in April 1945.

Fgst Nr.305082, 305083, 305084 and a fourth Jagdtiger constitute the last official recorded shipment of the war, and personnel from s.PzJg.Abt 653 had been sent to the Ni-Werk to pick them up. Feldwebel Rheinhold Schlabs, who commanded one of the Jagdtigers, was adamant that four were loaded onto railway cars and transported towards the front, arriving at Sankt Leonard am Forst, west of Vienna, on 1 May 1945. Some accounts state that the fourth vehicle was Fgst Nr.305085 but this would seem to contradict the accepted April production figure of seven Jagdtigers. It may be that the fourth vehicle was at Sankt Valentin undergoing repairs.

305085	April 1945	Henschel	Thought to have been either fitted with an 8.8cm gun or earmarked for one.
305086	April 1945	Henschel	Thought to have been either fitted with an 8.8cm gun or earmarked for one.
305087	April 1945	Henschel	Thought to have been either fitted with an 8.8cm gun or earmarked for one.
305088	April 1945	Henschel	Thought to have been either fitted with an 8.8cm gun or earmarked for one.
305089	May 1945	Henschel	Evidence for partially-assembed hull only. Probably never completed.
305090	May 1945	Henschel	Evidence for partially-assembled hull only. Probably never completed.
305091	May 1945	Henschel	Evidence for partially-assembled hull only. Probably never completed.
305092	May 1945	Henschel	Evidence for partially-assembled hull only. Probably never completed.

On 4 May 1945 men of s.PzJg.Abt 653, under the command of Leutnant Hans Knippenberg from the battalion's 1.Kompanie, destroyed the last eight Jagdtigers at the Ni-Werk plant in an effort to prevent their capture by the Soviets.

Photographed in the town of Bad Driburg, about 10 kilometres east of Paderborn, in April 1945, this Jadgtiger was probably disabled by its own crew. In other images of this vehicle the letter Y can be clearly seen on the hull side which has led many accounts to associate this Jagdtiger with 2.Kompanie, schwere Panzerjäger-Abteilung 512. But the battalion's first and second companies were lost in the fighting for the Ruhr Pocket, far to the west, and this vehicle probably operated with 3.Kompanie which is known to have suffered losses at nearby Beierfelde, Offensen, Sankt Andreasberg and Osterode am Harz.

The interior of Halle VII at the Sankt Valentin Nibelungenwerk after the Allied bombing raid of Monday, 16 October 1944. Fröhlich's account mentions that one Panzerwanne, or armoured hull, was so badly damaged that it was later scrapped. Note the holes bored in the hull side indicative of the Henschel suspension models. After the war Halle VII was dismantled by the firm of Wiener Brückenbau and the steel girders sent to Yugoslavia as reparations.

This photograph, taken in the grounds of the Nibelungenwerk in May 1945, shows that at least one of the last few production models was fitted with the 12.8cm gun. German accounts state that the last eight Jagdtigers were destroyed on 4 May 1945 to prevent them falling into the hands of the Soviets but as many as forty-five fully-assembled Pzkpfw IV tanks and a single operational Jagdtiger are reported to have been at Ni-Werk when the first Russians arrived five days later.

1

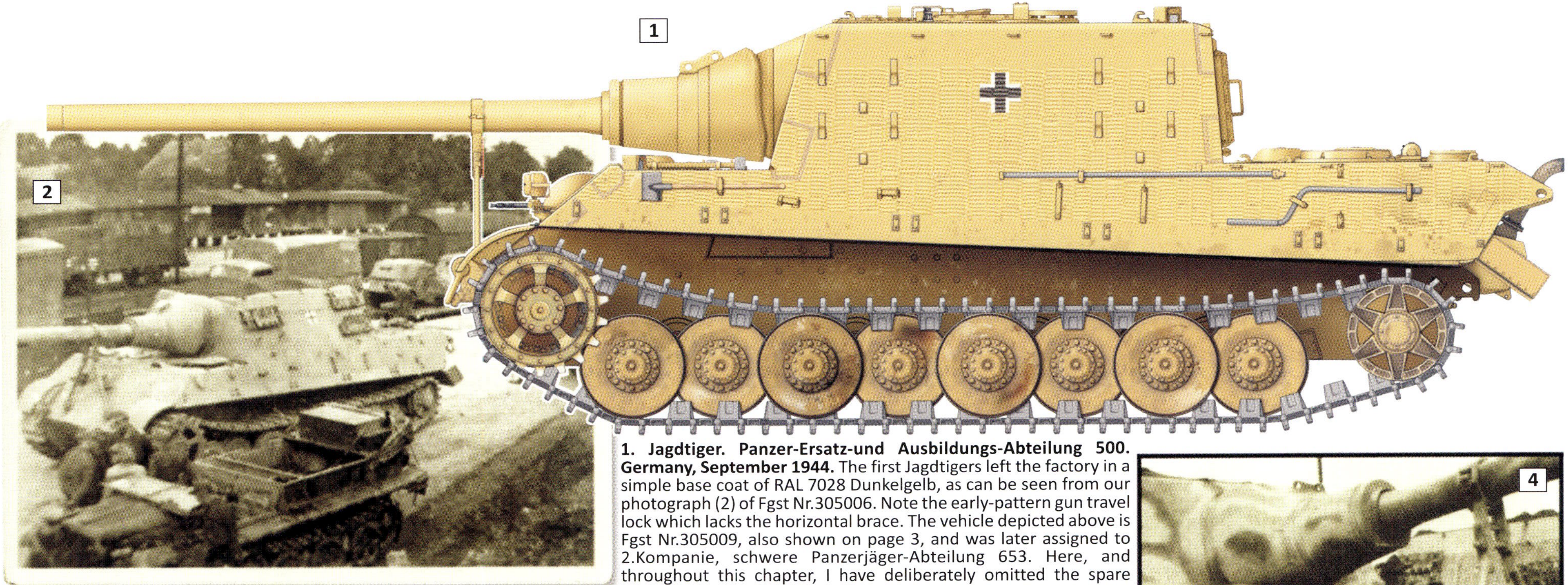

2

1. Jagdtiger. Panzer-Ersatz-und Ausbildungs-Abteilung 500. Germany, September 1944. The first Jagdtigers left the factory in a simple base coat of RAL 7028 Dunkelgelb, as can be seen from our photograph (2) of Fgst Nr.305006. Note the early-pattern gun travel lock which lacks the horizontal brace. The vehicle depicted above is Fgst Nr.305009, also shown on page 3, and was later assigned to 2.Kompanie, schwere Panzerjäger-Abteilung 653. Here, and throughout this chapter, I have deliberately omitted the spare tracks for the sake of clarity.

4

3

3. Jagdtiger. 1.Kompanie, schwere Panzerjäger-Abteilung 653. Eastern France, March 1945. This vehicle was one of the Jagdtigers that was assigned to Panzer-Ersatz-und Ausbildungs-Abteilung 500 for driver training but remained with 1.Kompanie when the battalion was sent to the front. A similar, almost identical, camouflage scheme can be seen on at least one other early arrival, Jagdtiger Fgst Nr.305010 of 3.Kompanie (4), and may have been applied to other vehicles of 1.Kompanie and 3.Kompanie which were equipped at the same time and trained together. It is very likely that the vehicle shown at left is Fgst Nr.305008.

1

2

1. Jagdtiger. 1.Kompanie, schwere Panzerjäger-Abteilung 653. Germany, April 1945. In mid-August 1944 armoured vehicle manufacturers were ordered to apply camouflage patterns as part of the production process. The so-called ambush scheme shown here, where contrasting dots or lozenges were applied to the patches of RAL 6003 Olivgrün and RAL 8017 Rotbraun, is generally accepted to have been short lived and the dots may have been a refinement that was abandoned due to the need to meet production targets. This Jagdtiger is probably Fgst Nr.305011, the first to be assembled with Zimmerit, and may also be the first example to leave the Nibelungenwerke with a factory-applied camouflage scheme. The company number is partly visible in our photograph (2) as is the 18-tooth drive sprocket and later-pattern gun travel lock. It was under the command of Leutnant Hans Knippenberg when it was abandoned near Schwetzingen in late March 1945.

3

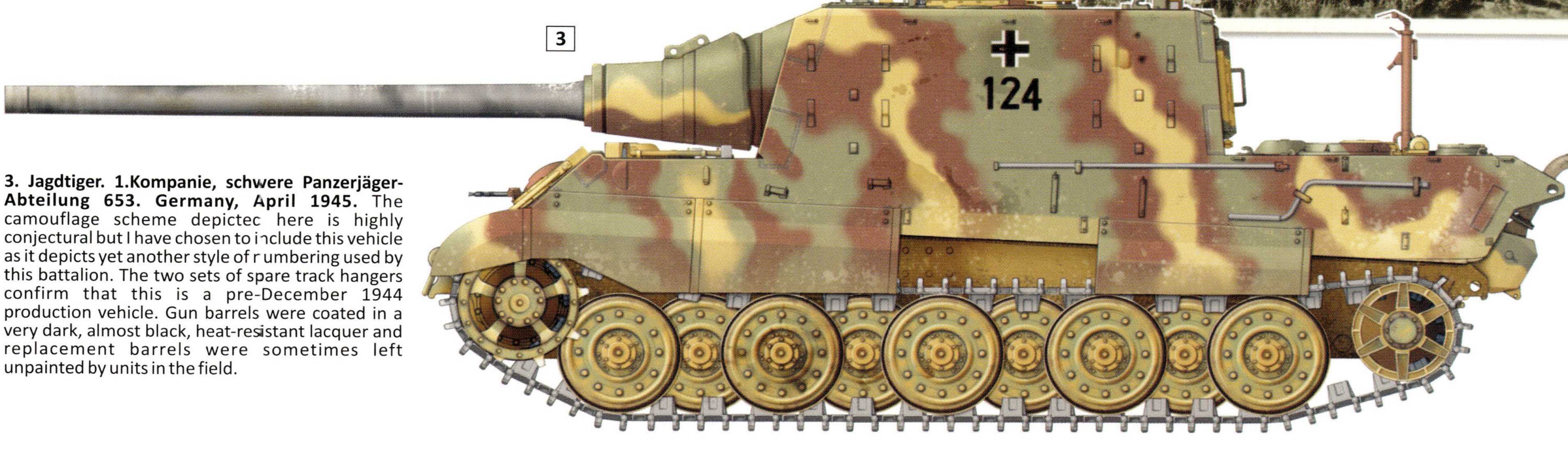

3. Jagdtiger. 1.Kompanie, schwere Panzerjäger-Abteilung 653. Germany, April 1945. The camouflage scheme depicted here is highly conjectural but I have chosen to include this vehicle as it depicts yet another style of numbering used by this battalion. The two sets of spare track hangers confirm that this is a pre-December 1944 production vehicle. Gun barrels were coated in a very dark, almost black, heat-resistant lacquer and replacement barrels were sometimes left unpainted by units in the field.

1

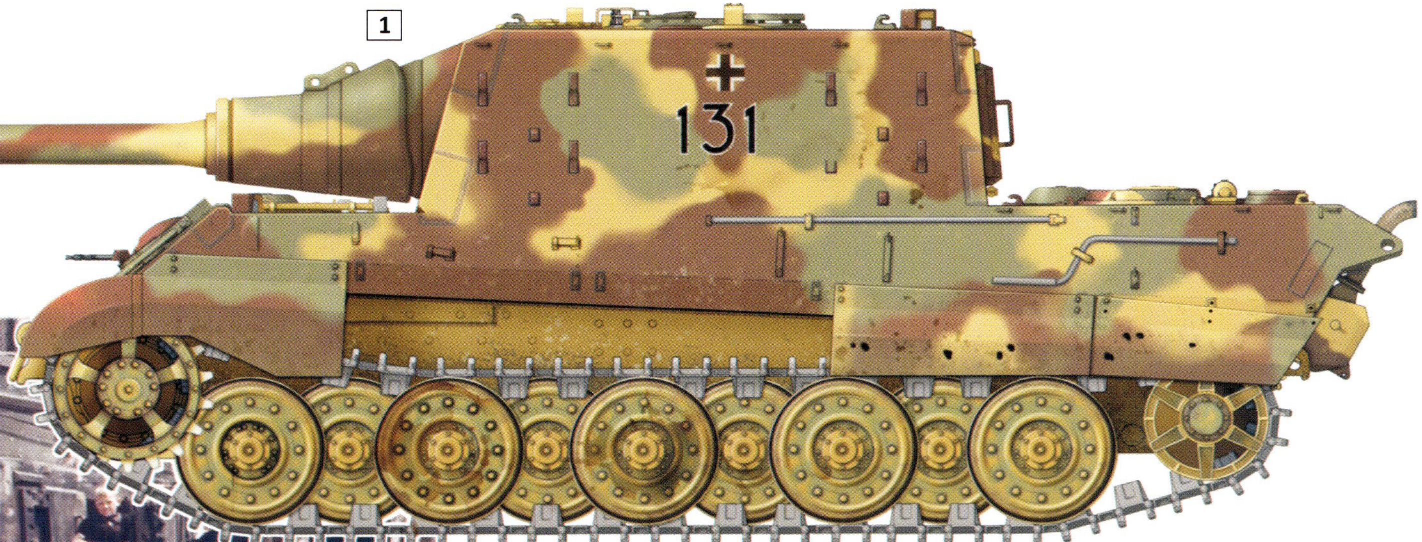

3

4

1 and 2. Jagdtiger. 1.Kompanie, schwere Panzerjäger-Abteilung 653. Germany, March 1945. Commanded by Oberfeldwebel Ferdinand Kinnberger, this vehicle was disabled by a close-range shot from a US Army Sherman tank and abandoned in the streets of Schwetzingen, west of Heidelberg, at the intersection of Heidelberger Strasse and Mannheimer Strasse. It has been suggested that this vehicle is Fgst Nr.305013 assembled in September 1944, the first Jagdtiger built with the Henschel suspension, but the many available photographs seem to show that it lacks the jack and jack block with their brackets. This suggests that it was manufactured in November 1944 when these fittings were discontinued and before the extra set of track hangers was introduced. The company number is similar in style to that applied to Jagdtiger 115, shown on the previous page, but the white outline has clear gaps as depicted here (3) and shown in our photograph (4). Note that the number on the right side has been very roughly applied.

2

1

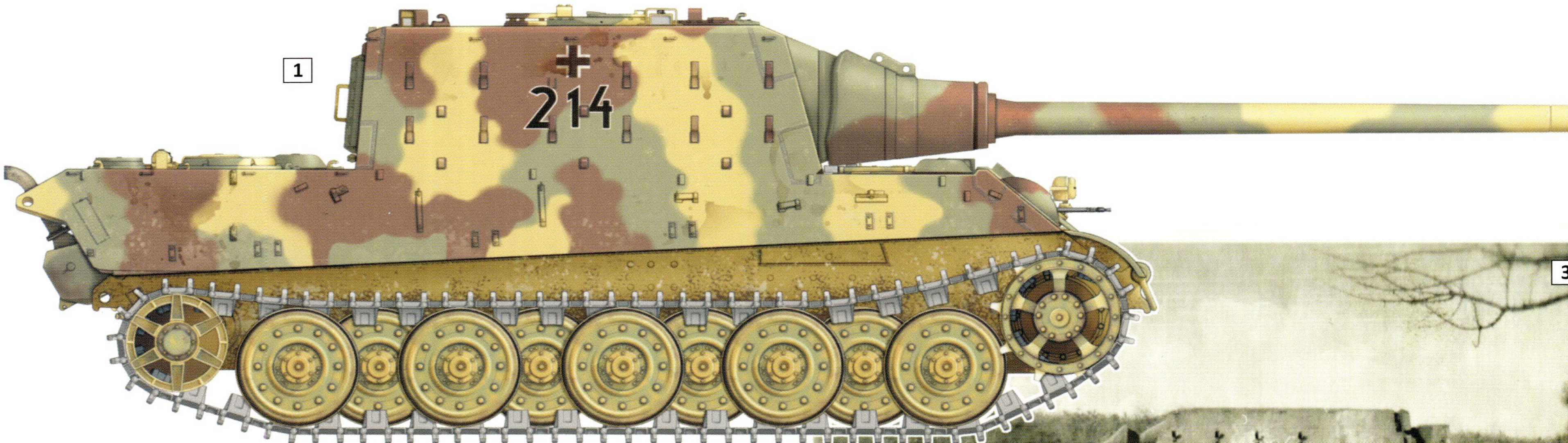

3

1 and 2. Jagdtiger. 2.Kompanie, schwere Panzerjäger-Abteilung 653. Germany, April 1945. This vehicle, Fgst Nr.305037, was lost near Nordheim, north of Stuttgart, on the road to Lauffen am Neckar while it was attempting to tow another Jagdtiger, possibly number 213. Assembled in December 1944, it has the three sets of track hangers on either side although the centre set appears to have been welded on over the company number and some of the metal hooks are missing. The camouflage pattern is also interesting as it has been neatly continued to the lower edges of the hull sides, suggesting that trackguards were never attached. The camouflage colours on the barrel and gun mantlet appear to have dark outlines in our photograph (3) but in other images they are solid blocks of colour as rendered here.

2

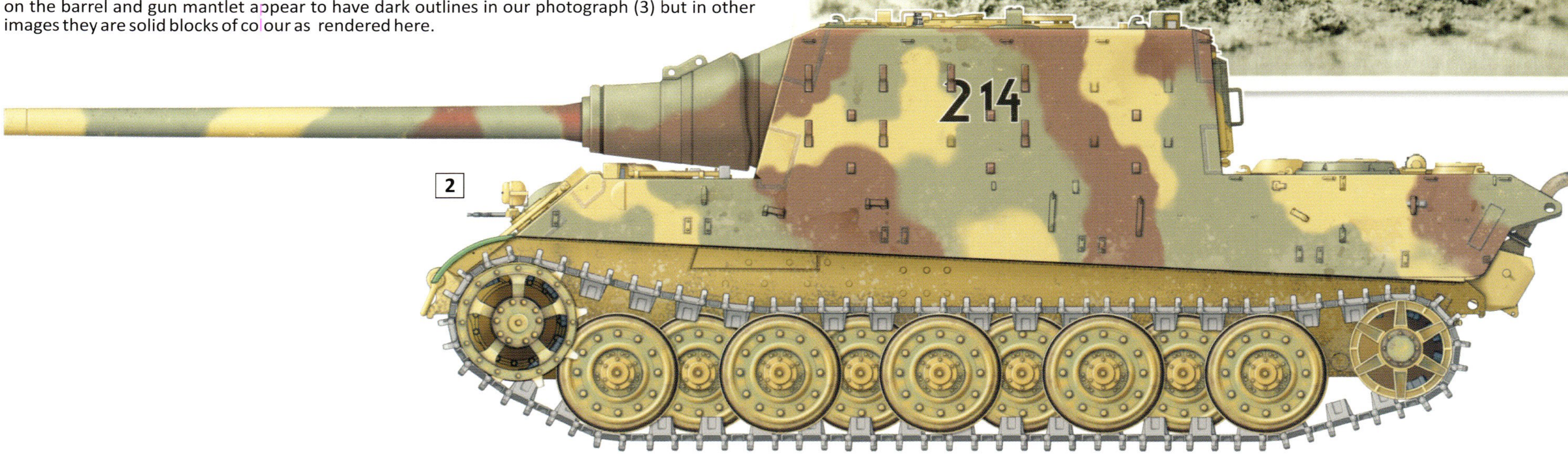

1

2

5

4

1. Jagdtiger. 2.Kompanie, schwere Panzerjäger-Abteilung 653. Germany, February 1945. Photographed outside the town of Hatzenweier, south of Stuttgart, this Jagdtiger was disabled by its own crew. Note that the centre set of track hangers appear to have been welded on over the company number (2) in the same manner as those fitted to Jagdtiger 214 shown on the previous page. This suggests that they were attached by the unit and not at the factory and if this is true the vehicle shown here may be a November 1944 production model. Note that the camouflage has been applied while the trackguards were in place, leaving the lower edge in its primer colour.

3

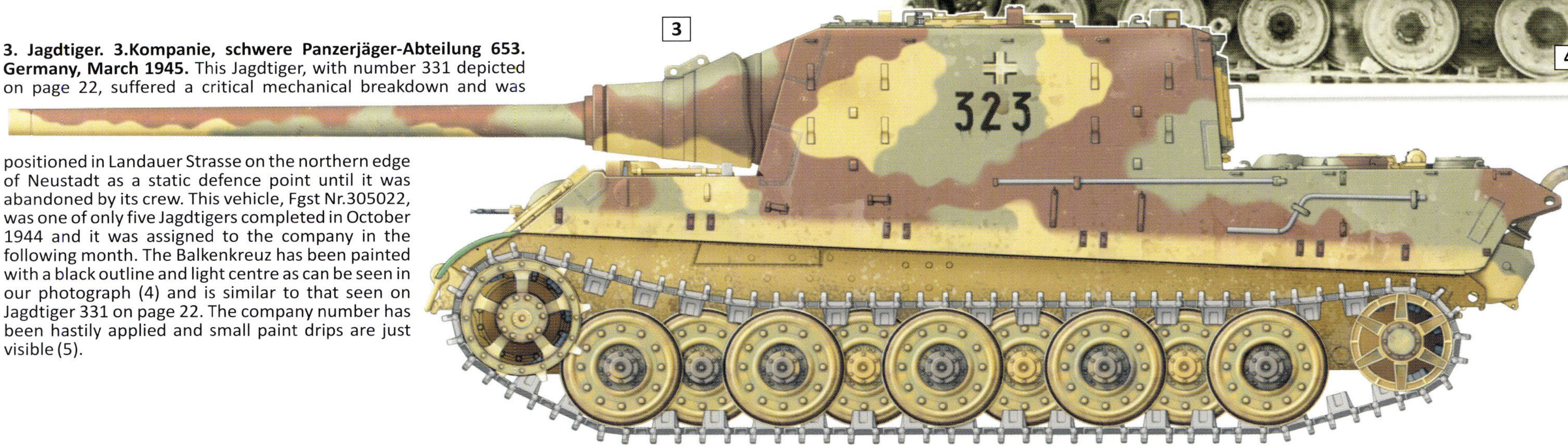

3. Jagdtiger. 3.Kompanie, schwere Panzerjäger-Abteilung 653. Germany, March 1945. This Jagdtiger, with number 331 depicted on page 22, suffered a critical mechanical breakdown and was positioned in Landauer Strasse on the northern edge of Neustadt as a static defence point until it was abandoned by its crew. This vehicle, Fgst Nr.305022, was one of only five Jagdtigers completed in October 1944 and it was assigned to the company in the following month. The Balkenkreuz has been painted with a black outline and light centre as can be seen in our photograph (4) and is similar to that seen on Jagdtiger 331 on page 22. The company number has been hastily applied and small paint drips are just visible (5).

1

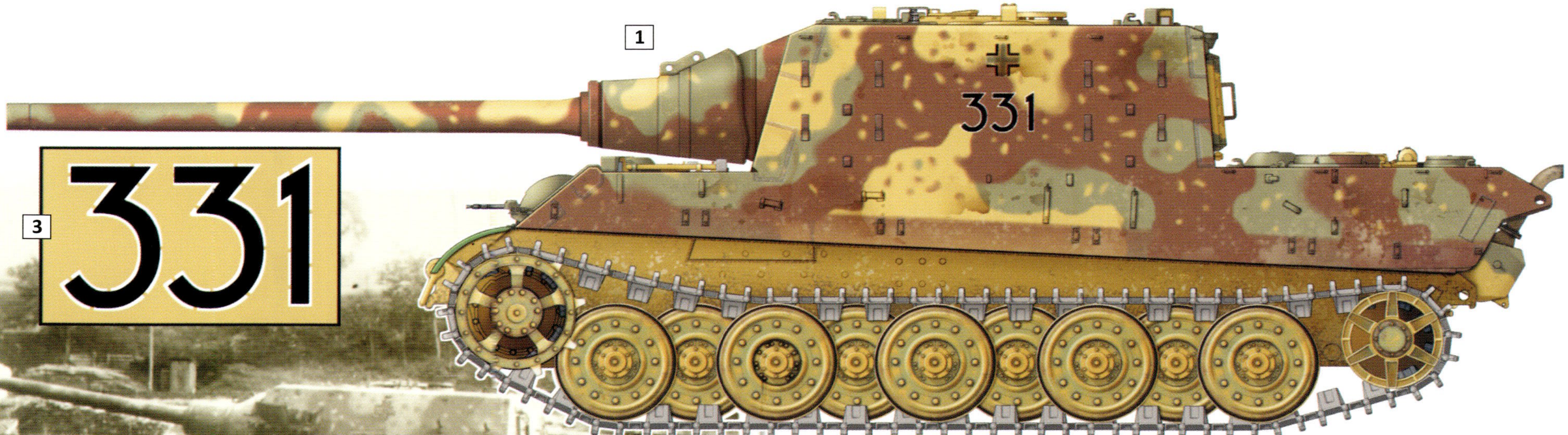

3

2

1. Jagdtiger. 3.Kompanie, schwere Panzerjäger-Abteilung 653. Germany, March 1945. Commanded by Leutnant Kaspar Göggerle, this Jagdtiger, and the vehicle shown on the previous page, were abandoned in Landauer Strasse in Neustadt, a small town north of Karlsruhe. The ambush camouflage scheme, which is clearly visible in our photograph (2), is similar to the pattern applied to Jagdtiger 115 depicted on page 18, albeit somewhat more comprehensive and developed. Although far from definitive, the camouflage suggests a production date of October 1944 and this vehicle may be Fgst Nr.305021. The company number is rendered in the same style as Jagdtiger 131 shown on page 19, almost certainly with the same stencil with its distinctive gaps (3). Note also the Balkenkreuz which has a black outline with light-coloured centre. After the war this vehicle was transported to the United States where it was put on display at the Army Ordnance Museum at Aberdeen, Maryland.

4

4. Jagdtiger. 3.Kompanie, schwere Panzerjäger-Abteilung 653. Germany, March 1945. Commanded by Feldwebel Heinz Telgmann, this Jagdtiger was disabled by American fighter-bombers near Morsbronn-les-Baines and abandoned by its crew. Note the darker outline of the Balkenkreuz which is rendered here as RAL 6003 Olivgrün although it could have been another colour of a similar shade. The company number has obviously been applied using the same stencil used for Jagdtiger 323 shown on the previous page but is completely different to that of Jagdtiger 331 although all three were probably both shipped to the battalion at roughly the same time.

1

2

3

5

1. Jagdtiger. 1.Kompanie, schwere Panzerjäger-Abteilung 512. Germany, April 1945. Commanded by Oberfähnrich, later Leutnant, Heinz Rondorf, this Jagdtiger was one of three that surrendered to US Army units at Iserlohn, south-east of Dortmund. The name, Der Haurgenaue (2), as rendered here was copied from a very clear photograph and it is often misspelt. It can be roughly translated as 'the exact one' or 'precisely that', but like many German phrases it is difficult to give a simple English version. It may have been a reference to Rondorf's marksmanship. The Balkenkreuz was also applied to the rear of the hull superstructure (3), probably on both sides as it was for the other vehicles in this group. Note the kill rings on the barrel of the main gun, made up of three thin and two thick bands. The numbering system used by this battalion, and this company in particular, is explained in detail on page 63.

4

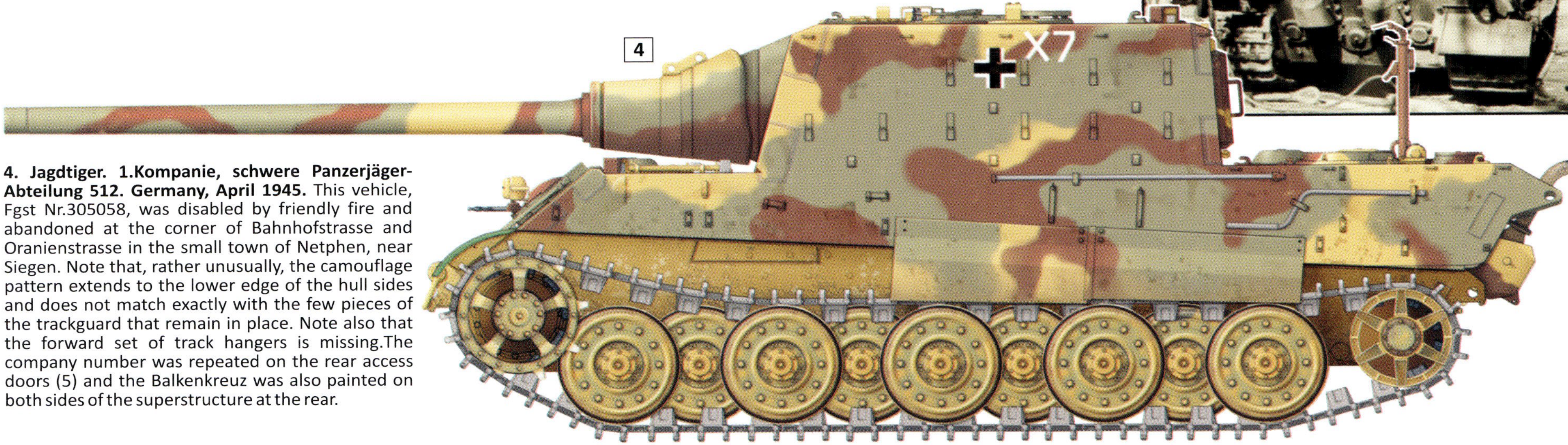

4. Jagdtiger. 1.Kompanie, schwere Panzerjäger-Abteilung 512. Germany, April 1945. This vehicle, Fgst Nr.305058, was disabled by friendly fire and abandoned at the corner of Bahnhofstrasse and Oranienstrasse in the small town of Netphen, near Siegen. Note that, rather unusually, the camouflage pattern extends to the lower edge of the hull sides and does not match exactly with the few pieces of the trackguard that remain in place. Note also that the forward set of track hangers is missing.The company number was repeated on the rear access doors (5) and the Balkenkreuz was also painted on both sides of the superstructure at the rear.

1

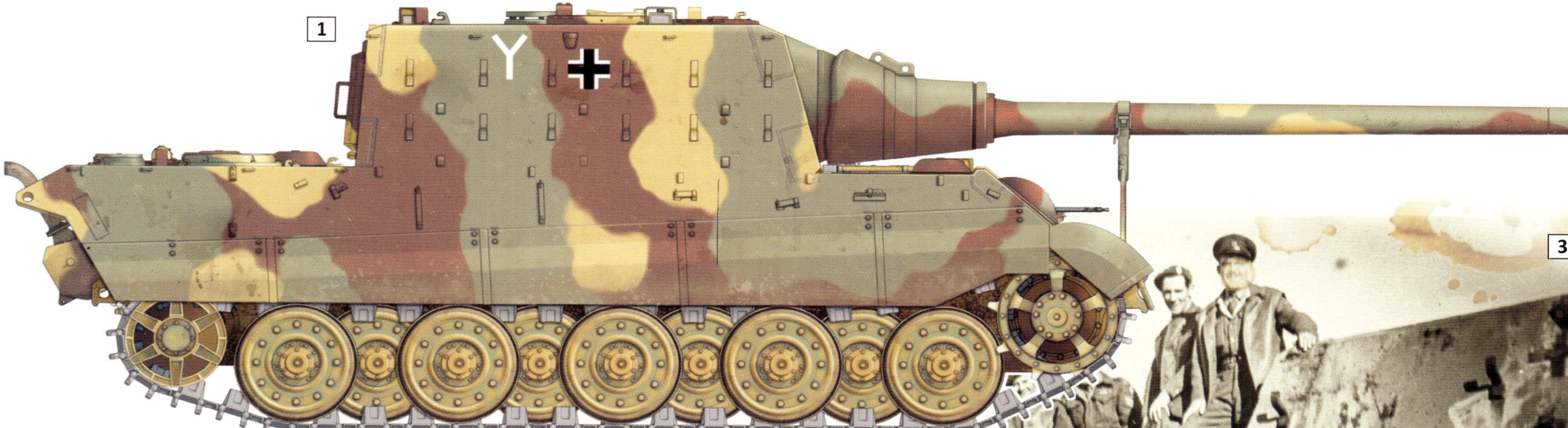

3

1. Jagdtiger. 3.Kompanie, schwere Panzerjäger-Abteilung 512. Germany, spring 1945. Many sources suggest that the white letter Y marking identifies this vehicle as a 2.Kompanie Jagdtiger and it has sometimes been assumed that it was commanded by Oberleutnant Otto Carius who led the company. Interestingly, in a post-war interview Herr Carius stated that he had no memory of the numbering system used by his company and that the Jagdtigers may not have been numbered at all. The location of the very clear photograph on which this illustration is based can be confirmed as Brakeler Strasse on the eastern side of Bad Driburg, a small town east of Paderborn. Given that the battalion's first and second companies were encircled in the Ruhr Pocket it seems far more likely that this is the Jagdtiger of 3.Kompanie which was reported lost in this area on 2 April 1945. Note that the front section of the mudguard has been fitted backwards. This vehicle is also depicted in the photograph on page 56.

2

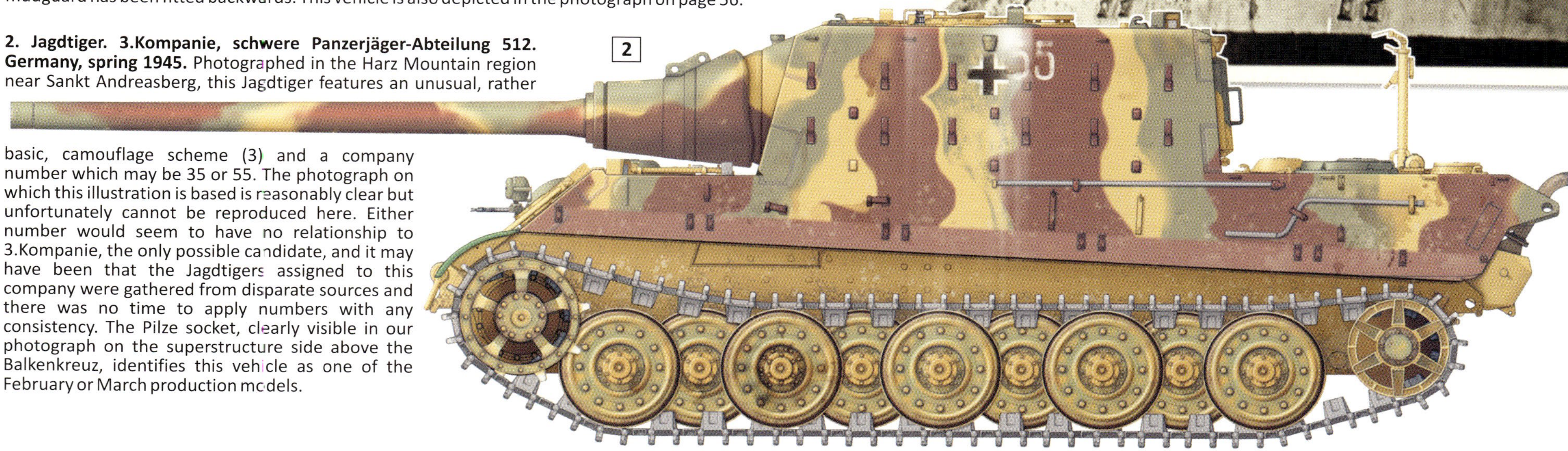

2. Jagdtiger. 3.Kompanie, schwere Panzerjäger-Abteilung 512. Germany, spring 1945. Photographed in the Harz Mountain region near Sankt Andreasberg, this Jagdtiger features an unusual, rather basic, camouflage scheme (3) and a company number which may be 35 or 55. The photograph on which this illustration is based is reasonably clear but unfortunately cannot be reproduced here. Either number would seem to have no relationship to 3.Kompanie, the only possible candidate, and it may have been that the Jagdtigers assigned to this company were gathered from disparate sources and there was no time to apply numbers with any consistency. The Pilze socket, clearly visible in our photograph on the superstructure side above the Balkenkreuz, identifies this vehicle as one of the February or March production models.

1

3

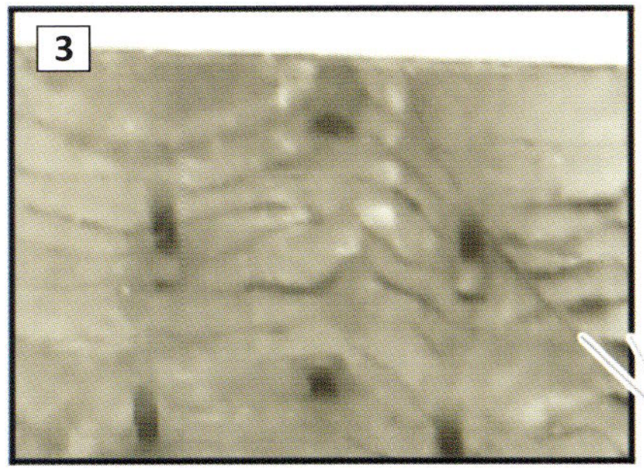

4

1. Jagdtiger. 2.Kompanie, schwere Panzerjäger-Abteilung 512. Germany, spring 1945. Photographed near Billmerich, about 10 kilometres east of Dortmund, this vehicle was attempting to tow another Jagdtiger when it was abandoned by the roadside. Manipulation of the available colour image shows what appears to be a 2.Kompanie number and the company did operate in this area as part of a Kampfgruppe made up of 116.Panzer-Division and 190.Infanterie-Division.

2. Jagdtiger. 3.Kompanie, schwere Panzerjäger-Abteilung 512. Germany, April 1945. Photographed outside the village of Kamschlacken in the Harz Mountains, this vehicle was the company's last surviving Jagdtiger and is almost certainly the vehicle commanded by Feldwebel Otto Arnold and referred to several times in official orders as Jagdtiger number 321. Curiously, there is a marking on the superstructure side which could be I1 or 11 (3). Note the Pilze sockets on the side, rear and front and the armoured cover for the command radio antenna which is clearly visible in our photograph (4). Although not shown here, this vehicle was fitted with the narrow Verladekette transport tracks.

2

1

2

4

1. Jagdtiger. Kampfgruppe SS-Panzer-Regiment 1. Austria, May 1945. This vehicle, Fgst Nr.305083, was one of four Jagdtigers picked up from the Nibelungenwerk by men of schwere Panzerjäger-Abteilung 653 in April 1945 as mentioned in the unit histories. It was captured intact by the Soviets and tested extensively and it seems unlikely that the Russians would have repainted this Jagdtiger in such an elaborate scheme, including the accurately sized and placed Balkenkreuz markings (2). It would therefore be safe to assume that this is how the vehicle appeared when it left the assembly line and it presents a fascinating picture of how the factory-applied camouflage schemes had evolved. The wheels may have also been painted, as they certainly were on the Panthers that were manufactured at this time.

3

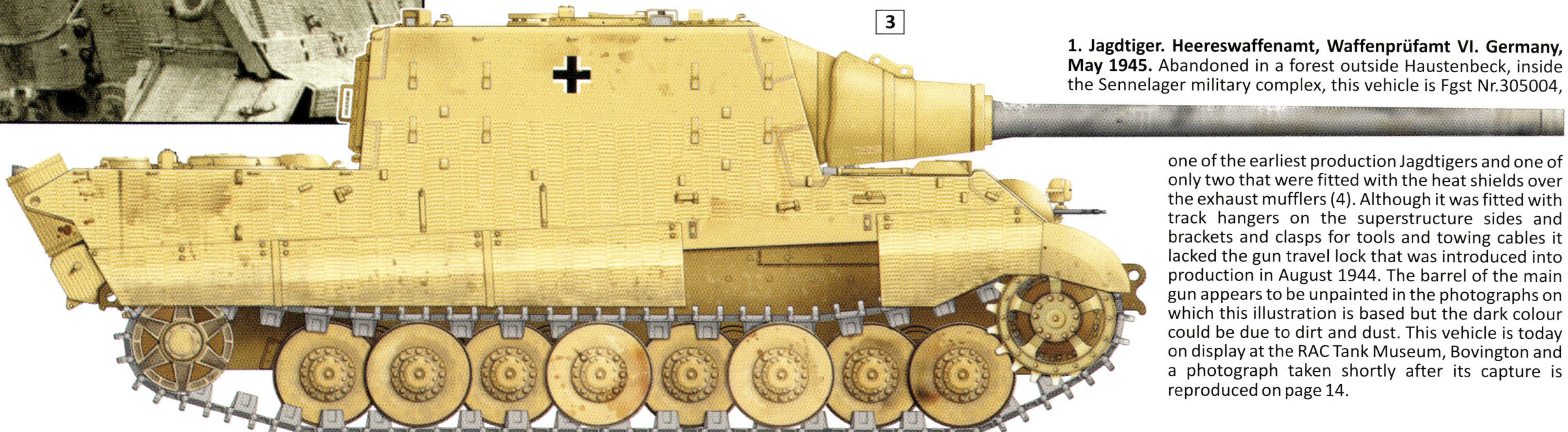

1. Jagdtiger. Heereswaffenamt, Waffenprüfamt VI. Germany, May 1945. Abandoned in a forest outside Haustenbeck, inside the Sennelager military complex, this vehicle is Fgst Nr.305004, one of the earliest production Jagdtigers and one of only two that were fitted with the heat shields over the exhaust mufflers (4). Although it was fitted with track hangers on the superstructure sides and brackets and clasps for tools and towing cables it lacked the gun travel lock that was introduced into production in August 1944. The barrel of the main gun appears to be unpainted in the photographs on which this illustration is based but the dark colour could be due to dirt and dust. This vehicle is today on display at the RAC Tank Museum, Bovington and a photograph taken shortly after its capture is reproduced on page 14.

JAGDTIGER

LAST PRODUCTION RUN

1/16 SCALE

KLAUS DELVENTHAL

Klaus' model represents one of the very last production vehicles to leave the Nibelungenwerk assembly line. The basis for the project was the Trumpeter 1/16 scale Jagdtiger but as the kit represents an earlier production version a number of parts needed to be scratch-built including the ribbed front fenders, the saw and shovel and their brackets, the extra Pilze and track hangers and the armoured cover for the insulator of the Sternantenne D command aerial. In addition Klaus reworked all the weld seams and added a handle over the rear door, additional armour for air filters, a hold-open catch for the engine deck and a bolted-on plate over the deep-fording intake.

Above: Details of Klaus' completed Jagdtiger including the late-production ribbed front fenders, the large towing coupling that was probably only fitted to the last few vehicles, the exhaust muffler, with appropriated jack, and the photo-etched engine grills from Aber and Tetra fire extinguisher from AFV Modeller.

Both images on this page show the completed model after painting and weathering. Klaus first used Tamiya Surface Primer and then created the camouflage scheme with Lifecolor Acrylic Hobby Colours. The pattern is a faithful representation of that seen in several views of Fgst Nr.305083, the Jagdtiger captured intact at Strenberg in May 1945. Klaus then gave the model several washes with AK Interactice Oilcolours and a coating of mud and dust using MIG Productions Pigments.

JAGDTIGER

1.Komp/s.PzJg Abt 653

1/35 SCALE

MUMU ZHENG

Mumu Zheng's replica is based on the Blitz by Takom 1/35 scale kit, marketed as Early/Late Production 2 in 1 model. The camouflage scheme and markings are based on the 1.Kompanie, schwere Panzerjäger-Abteilung 653 vehicle that was abandoned in Schwetzingen in March 1945. Shown at the centre of the page are a number of details including the superstructure track hangers and spare track, the Fliegerdrehstütze 36 anti-aircraft MG 42 mount and the Henschel-designed rear idler.

Another view of the completed model. Note that a number of the track hanger hooks are missing as they frequently were on the actual vehicles. Note also the brackets for the 20-ton jack, visible just below the armoured covers for the exhaust mufflers, and the wooden jack block on the right side of the hull rear plate. These features were dropped from production in late October 1944.

Further views of the finished model showing the superstructure and hull upper surfaces. Note the rangefinding periscope and its hinged cover in front of the commander's position.

Details of the engine deck showing the photo-etched mesh covers for the air intakes. Note the position of the base for the Fliegerdrehstütze 36.

JAGDTIGER

3.Komp/s.PzJg Abt 653

1/35 SCALE

KEVIN D. POTTS

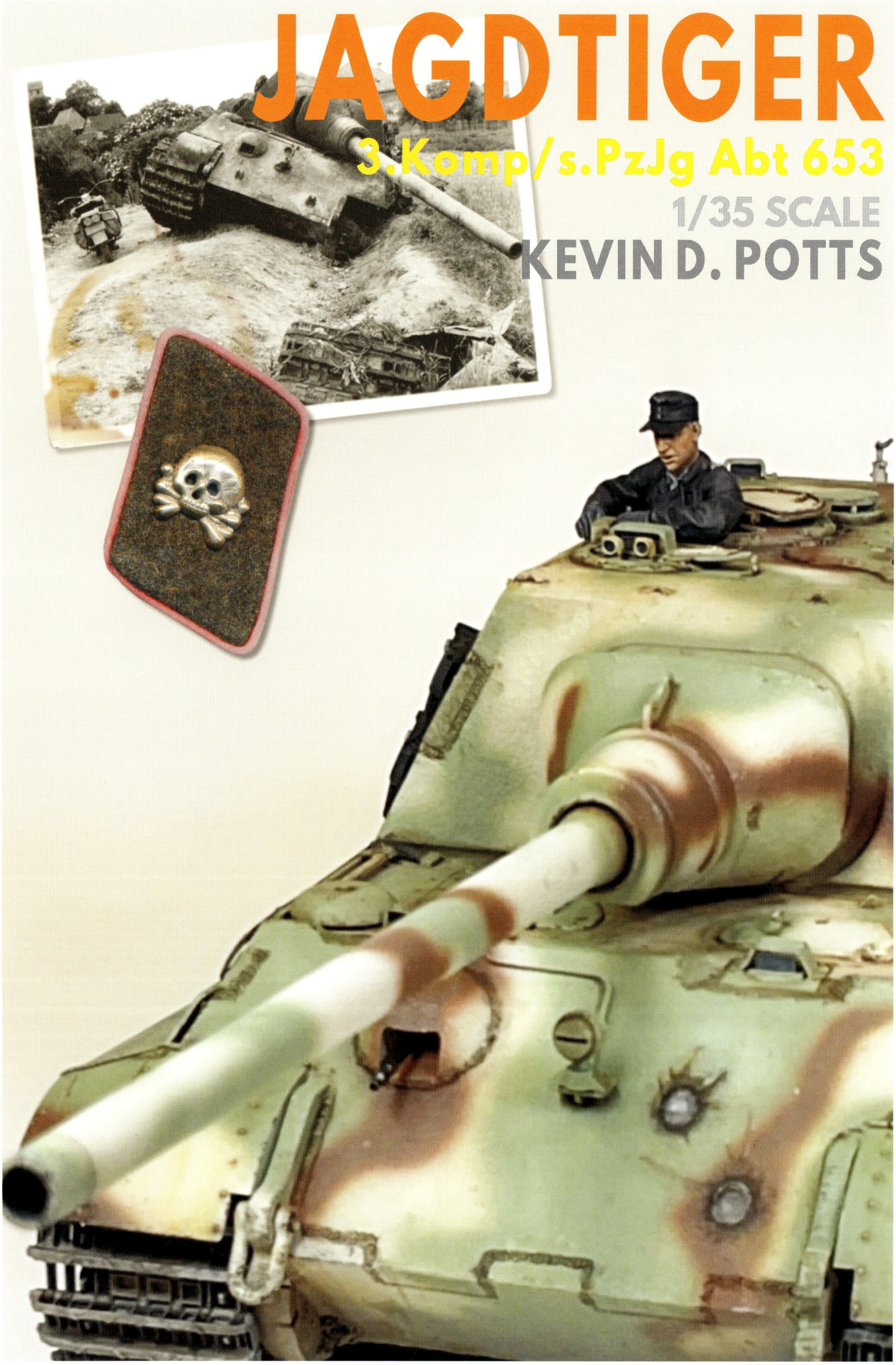

Kevin's model is based on the Blitz by Takom 1/35 scale Porsche Production Type kit and finished to represent a Jagdtiger of 3.Kompanie, schwere Panzerjäger-Abteilung 653. All the weld seams were scraped off and new versions added using Magic Sculpt and the tie downs and tool clamps and clasps were also replaced. The trackguards are those supplied with the cut but some sections were removed and all were drastically thinned.

Below: Details of the finished model including the Kugelblende machine-gun mount and Bosch headlight, the driver's hatch and periscope, the superstructure roof with ventilator/fume extractor cover and periscope, the commander in his hatch and another view of the fume extractor.

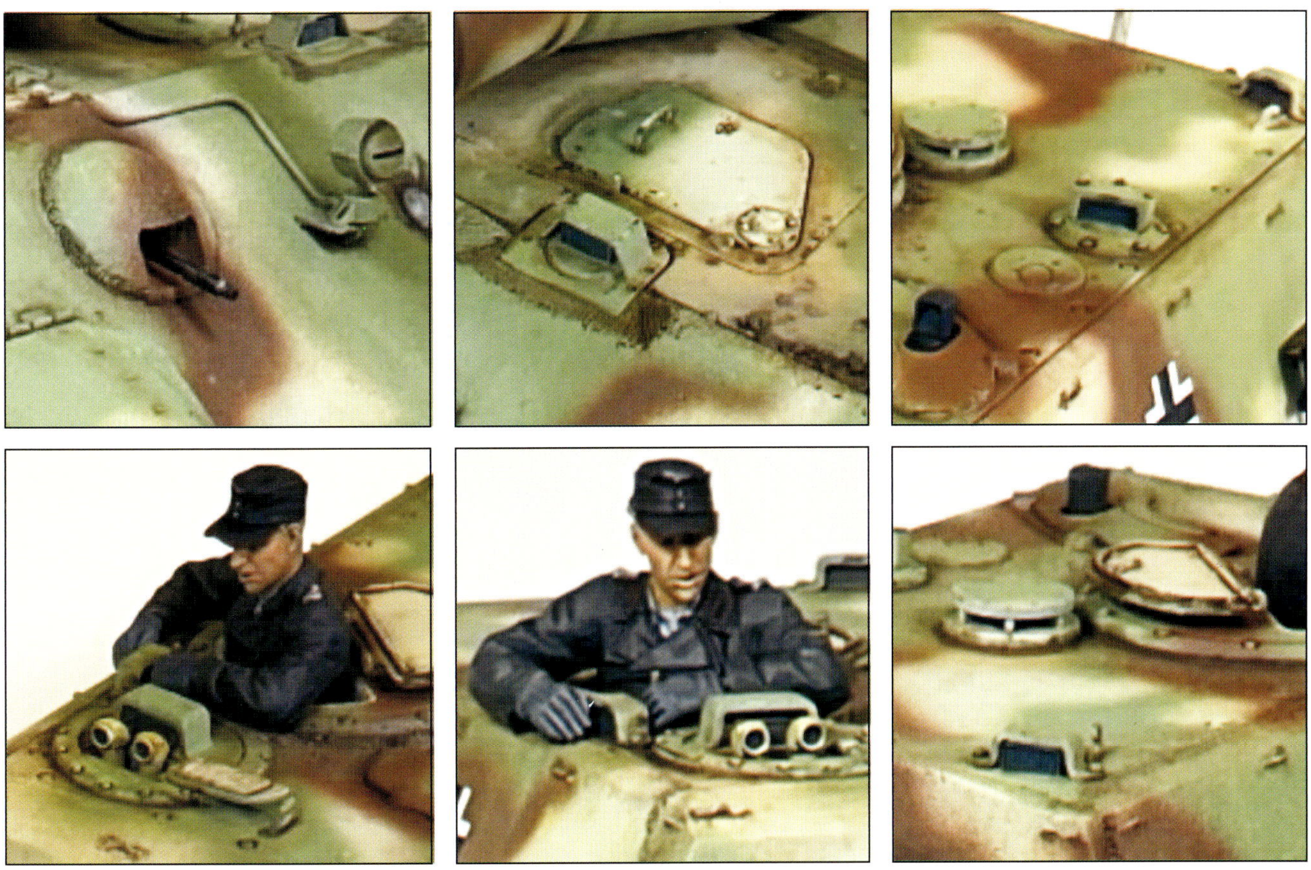

Once construction was complete Kevin painted the entire model in Ammo Black Primer. The soft-edge camouflage scheme was created with AK Interactive Dark Yellow and Pale Green with Vallejo Flat Brown. Each shade had various highlights added before moving on to the next colour. The shell damage on the hull glacis was made by gouging the flat plastic surface with a drill bit and then filling the hole with Squadron Gray Putty. While soft, the putty was shaped with the end of a paint brush to get the right amount of metal damage. Note the detailed TREX resin tracks which Kevin used to replaced those from the kit.

When the final coat was dry Kevin began weathering the hull and suspension beginning with a mixture of dirt from his driveway and VMS Sand and Ballast Freeze. Next, Vallejo Dark German Gray was used to build up paint chips followed by several layers of mud coloured paint based on AK Weathering Enamels and artists' oils.

JAGDTIGER

3.Komp/s.PzJg

1/35 SCALE

YOHHEN MYURĀ

Yohhen's model is the Tamiya 1/35 scale kit built using the parts supplied by the manufacturer with some minor alterations and scratch-built details. At the bottom of the page are details of the partly-completed model showing the texture that Yohhen was able to create with a mixture of talcum powder, Tamiya putty and lacquer thinner. The weld seams were made using Tamiya putty and the hinges for the fender are pieces of plastic card.

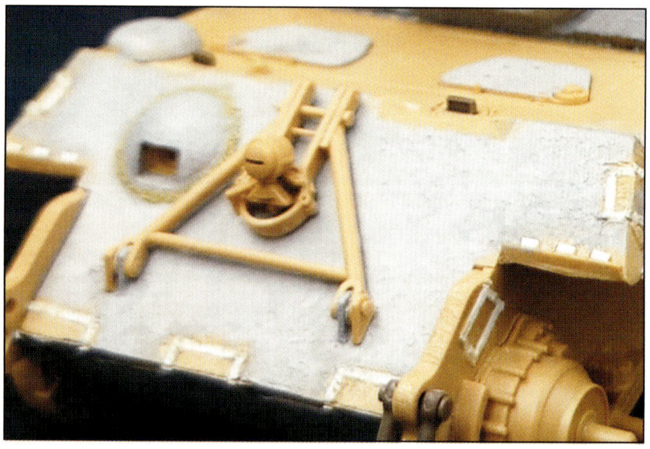

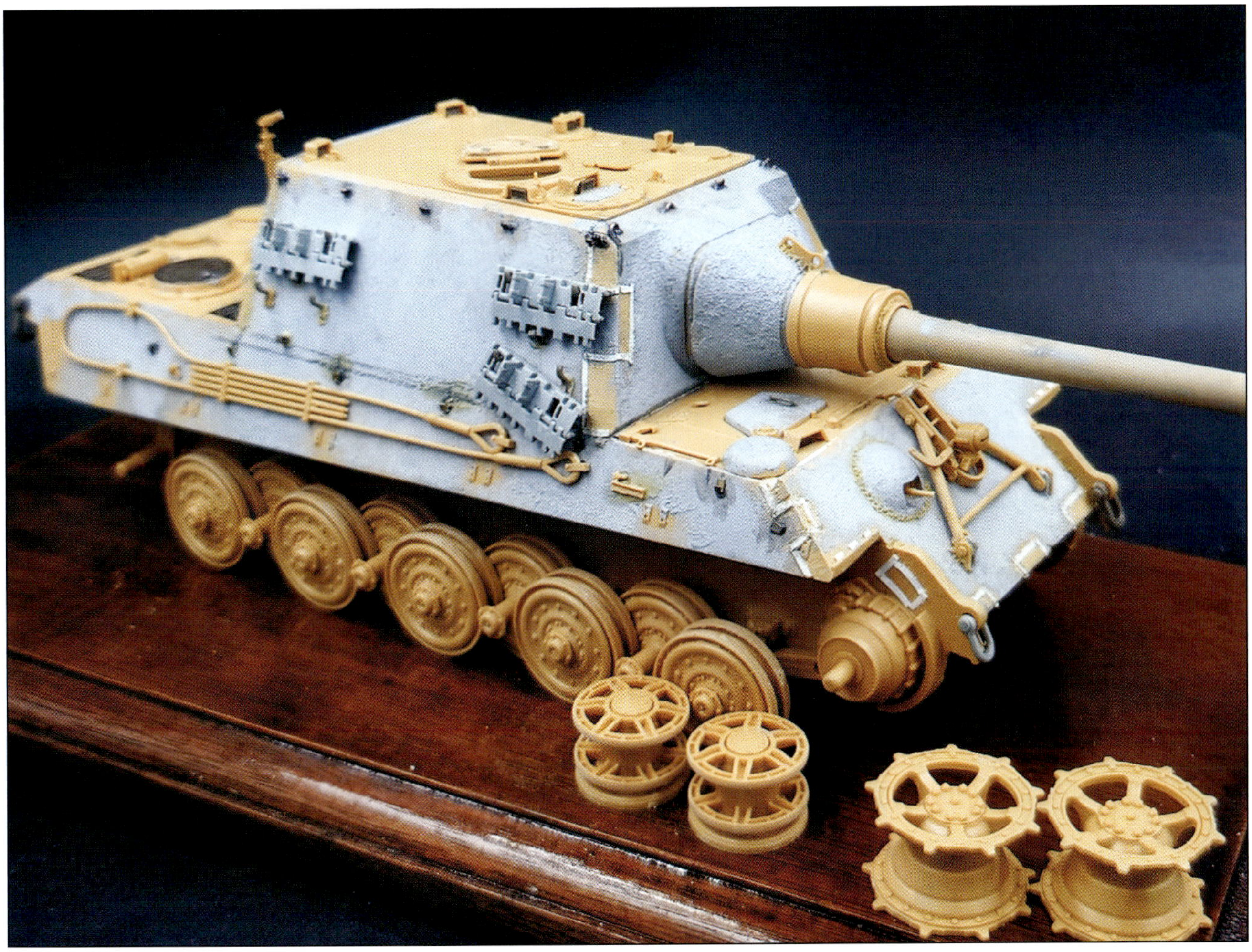

Above: The fully-assembled hull and superstructure. The road wheels, drive sprocket and rear idler were constructed and dry fit before painting. Below: The model was painted using Tamiya acrylics while the markings are a mixture of kit transfers and hand painting

The completed model on its base. The tracks are the soft plastic version supplied with the kit and were weathered with a mixture of dirt, powder and acrylic colour after they were fitted to the suspension. Note the folded jacket hanging from the machine-gun mount and the detailed fire extinguisher.

The commander figure is from Alpine Miniatures. This US-based company was founded by well-known modeller Taesung Harmms and offers an extensive range of highly-detailed and anatomically superb figures.

A rear view of Yohhen's completed model with detail shots below. These include the folded gun travel lock and Bosch headlight, long-handled shovel and mallet, photo-etched air intake covers, superstructure roof with gunner's periscope and opened commander's hatch, and the double lifting eye welded to the top of the gun mantlet.

JAGDTIGER

3.Komp/s.PzJg Abt 653

1/35 SCALE

JAMES NIGEL PANGANIBAN

Regular readers of this series will be familiar with James Nigel's work and the model shown here, built to his usual high standards, is based on the Tamiya 1/35 scale kit. The model comes complete with photo-etched engine air intake grills and a choice of either link-and-length or vinyl one-piece tracks.

Below: Detail shots of the Jagdtiger's massive gun mantlet and the front plate of the superstructure showing the cast texture of each.

James Nigel's finished model represents Fgst Nr.305020, a September production vehicle that was later assigned to 3.Kompanie, schwere Panzerjäger-Abteilung 653. The intricate camouflage scheme was painted using Tamiya Dark Yellow II, Dark Green II, and Red Brown II. The markings are those supplied with the kit. The final weathering was achieved using a mix of enamel paints, artists' oils and various specialised weathering products from Vallejo and AK Interactive but was kept to a minimum.

Below: Details of James Nigel's finished model showing the gun travel lock, rear access doors and exhaust mufflers. Note the use of the correct Fahrgestellnummer on the hull glacis just below the travel lock.

As one of the most famous armoured fighting vehicles produced during the Second World War it should come as no surprise that model replicas of the Jagdtiger have been produced in sizes ranging from 1/285 scale wargames tanks to massive 1/6 scale radio-controlled vehicles, together with a bewildering array of accessories. Unfortunately there are far too many to list here individually and, as with the other books in this series, I have attempted to concentrate on the kits offered by the major manufacturers in the most popular modelling scales. As many of the currently available accessory sets are offered by the companies showcased in earlier titles from this series, including *Tank Craft 25: Tiger I & Tiger II Tanks German Army and Waffen-SS, Normandy Campaign 1944* and *Tank Craft 31: Tiger I and Tiger II Tanks German Army and Waffen-SS, The Last Battles in the East, 1945,* I have chosen to expand the Model Showcase section of this book and reduce the number of pages that would normally make up the Modelling Products chapter. As always, an index of manufacturers, with their contact details, can be found on page 64.

DRAGON MODELS

Based in Hong Kong, Dragon Models was one of the first of the major manufacturers to offer accurate replicas in 1/35 scale and their kits are today something of a benchmark. The company's small scale releases are some of the best available. In 1/35 scale the company's current catalogue contains a Jagdtiger with Porsche suspension, both with and without Zimmerit texture, a Henschel suspension version and a Henschel model with a number of late production features. These have been re-boxed several times and upgraded with new parts since the first kit was released in 1995. The 1/35 scale model offered by Revell in 1998 was in fact a re-boxed version of Dragon Models' Henschel Production Type kit. In 2004 the company released a 1/72 scale version with the Porsche suspension and this was quickly followed by a similar kit with Zimmerit texture and in 2010 a Henschel suspension version.

At right: Dragon Models' 1/72 scale Porsche suspension Jagdtiger with Zimmerit kit completed with the upgrade set from E.T. Model. At left: Details of the 1/35 scale Jagdtiger model with the later-style drive sprocket and commander's hatch.

TRUMPETER

First released in 2016, the company's 1/16 scale Jagdtiger includes interior detail, photo-etched parts, a turned aluminium barrel, metal accessories and several marking options. The company also offers a range of 1/72 scale Jagdtiger models with both the Porsche and Henschel suspension units. Klaus Delventhal's highly detailed and extensively researched replica based on the Trumpeter 1/16 scale model can be seen in the Model Showcase section of this book.

Above: Trumpeter's 1/72 scale Porsche suspension Jagdtiger. The first model in this series was released in 2008 and there is also a Henschel suspension and an 8.8cm-armed version available. These are reasonably detailed kits and feature a number of small parts for the suspension and hull fitting. At right: One of the pre-painted and assembled kits from Easy Model which is essentially the Trumpeter product.

TAMIYA INCORPORATED

This Japanese company is currently the largest producer of scale model kits in the world, releasing their first 1/35 scale version of the Jagdtiger in 1967. This kit was upgraded and re-boxed a number of times until 2008 when a completely new version was released. This model was a great improvement on the earlier offering and featured photo-etched brass details, anatomically correct crew figures and accurate marking options for five different vehicles. In 2010 Tamiya released a 1/48 scale Jagdtiger model which is basically a scaled-down version of their 1/35 scale kit. All the company's Jagdtigers feature the Henschel suspension. The 1/35 scale model is also available as part of the Girls und Panzer franchise.

Far right, above: Tamiya's 1/48 scale Jagdtiger. At right: The 1/35 scale kit built with the photo-etched brass detail set from Eduard Model Accessories.

ACADEMY PLASTIC MODEL

This Korean company first released a 1/48 scale model of the Jagdtiger in the early 1980s which was re-released a few years later as a motorised version. It was a fairly rudimentary replica, even by the standards of the day, and it appears to have been dropped from the company's catalogue.

Above: Academy's third re-boxing of the 1/48 scale motorised Jagdtiger model. At right: The same kit was re-released in the late 1990s with different marking options.

TAKOM

One of the newest model manufacturers, this company started operating in Hong Kong in 2013 as Takom World and shortly afterwards expanded to open a plant in mainland China where all the models are currently produced. First released in 2019, the company's 1/35 scale Jagdtiger with Henschel suspension was quickly followed by a Porsche suspension version, a model with the option for the installation of the 8.8cm gun and a Porsche variant with Zimmerit and exhaust muffler covers, features applicable to a small number of production vehicles. All these kits were produced under the company's Blitz label.

The highly-detailed 1/35 scale Henschel suspension Jagdtiger from Takom released in the Blitz range. Below: Details of the same model showing the drive sprocket fitting, superstructure roof, tracks and engine deck.

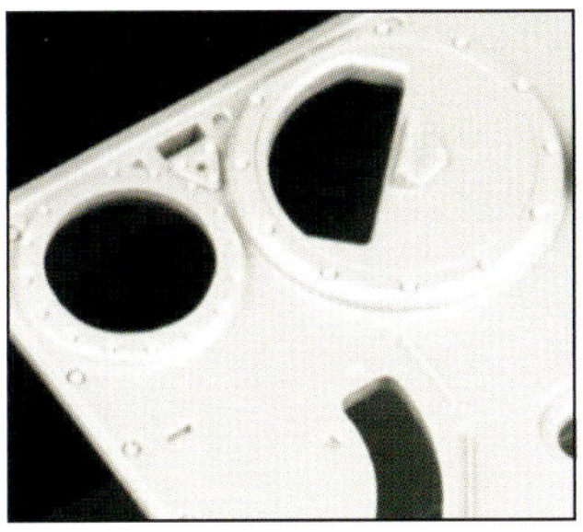

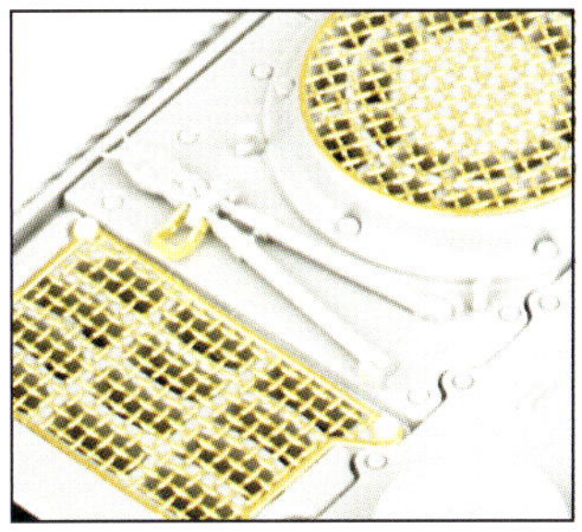

HOBBY BOSS

In 2022 the company released a 1/35 scale Jagdtiger with Porsche suspension followed by a Henschel version in the following year. I should mention that although the box illustration for the Porsche suspension model shows a vehicle coated with Zimmerit anti-magnetic mine paste this is not included in the kit. Hobby Boss and Trumpeter are in fact the same company.

Below: Some of the 980 parts that are included in the Hobby Boss 1/35 scale Henschel Jagdtiger kit.

ITALERI AND ESCI

In addition to the 1/72 scale re-boxed ESCI Jagdtiger mentioned below, this Italian company produces a 1/56 scale version of the Henschel suspension variant. This model features simplified construction, although it is reasonably accurate, is compatible with 28mm wargames figures and is part of an expanding range. Interestingly named for the Ente Scambi Coloniali Internazionali, or Colonial Trade Exchange, ESCI was founded in 1930 and began producing plastic model kits in the early 1960s. The company was responsible for several important innovations including the first hard plastic link-and-length tracks which ESCI introduced in 1985. The 1/72 scale Jagdtiger was first released in 1974 and re-boxed many times, including versions offered by Hasegawa, Revell and Humbrol. In 1988 the Jagdtiger was re-released with the new hard plastic tracks and this version is still offered by Italeri which bought many of the company's moulds when ESCI closed its doors in 1993.

Above: The ESCI 1/72 scale Jagtiger shown here as the 1988 re-release with hard plastic tracks. At right: Italeri's 1/72 scale Jagdtiger which is the ESCI model with new marking options. Below: Details common to both examples showing the upper superstructure with hatches and interlocking armour plates, the rear engine deck and the Henschel road wheels with early drive sprocket and later-pattern rear idler.

OTHER MANUFACTURERS

A number of other companies offered models of the Jagdtiger and although many of these are long out of production they are popular with modellers and collectors alike and should at least be mentioned.

Bandai. In the 1970s this Japanese company offered what were probably the most detailed armour models in 1/48 scale which are still sought by collectors today. The company's Jagdtiger was available only as the Henschel version and was re-boxed a number of times. It was also released under the Mitsuwa Models label until the late 1980s, again in differently-boxed versions, and by Fuman which simply changed the Bandai company name and retained the original box art.

Fujimi. One of the few companies, other than Airfix, to produce armour kits in 1/76 scale, this Japanese company released their first Jagdtiger model in 1971. Although best described as basic, these models were reasonably accurate and easy to assemble and contained some interior parts, notable the engine. The 1990 re-released model, which was in fact the Nitto kit, featured photo-etched air intake grills and spare tracks in white metal.

Nichimo. The company's catalogue included static and motorised Jagdtiger kits in 1/35 scale, both with the Henschel suspension. These were first released in 1975 and upgraded several times but seem to have been dropped from the catalogue and re-released by Korean firm Hap Dong Tech in the late 1980s.

Aoshima. In 2010 the company released a remotely-controlled 1/48 scale model based on the Jagdtiger with Henschel suspension. It was upgraded with new parts in 2020 but seems to have been discontinued shortly after that. It is still available from a number of retailers.

Otaki. In 1974 this Japanese company offered a 1/35 scale Jagdtiger complete with remotely-controlled motor. The company ceased trading in 1986 but examples of the Jagdtiger model can be found at online auction houses, often at high prices.

From February 1944, when the first production series Jagdtiger was accepted, until May 1945, when the assembly plant was overrun by units of the Red Army, these heavy armoured vehicles were subject to almost constant change and upgrading. These modifications were sometimes introduced as economy measures but also as a means of increasing the Jagdtiger's combat efficiency and to remedy a number of technical problems, many of which bedevilled the design throughout its existence. It should be mentioned that many available records are ambiguous, or even contradictory, while others simply did not survive the war and the picture presented here has been compiled from a number of sources including photographic evidence. Therefore, this section is not intended to be a definitive history of the development of the Jagdtiger but will hopefully serve as a useful spotter's guide and also explain some of the terminology which many readers may find confusing. As the Jagdtiger was essentially based on the suspension and hull of the Tiger II, a number of modifications incorporated into that programme were applied to the new design. Where this is so I have made mention of it but readers who wish to examine the alterations made during production of the Tiger II will find that they are discussed in detail in *TankCraft 13: Tiger I and Tiger II Tanks: German Army and Waffen-SS, The Last Battles in the West, 1945, TankCraft 25: Tiger I and Tiger II Tanks: German Army and Waffen-SS, Normandy Campaign 1944* and *TankCraft 31: Tiger I* and *Tiger II Tanks: German Army and Waffen-SS, The Last Battles in the East, 1945.* As explained earlier, each vehicle was identified by a unique chassis number, or Fahrgestellnummer, here abbreviated as Fgst Nr.

February 1943. At a conference held at the beginning of the month, the need for a super-heavy Sturmgeschütz, or assault gun, ideally armed with the 12.8cm gun, was raised by the representatives of the army's Wa Prüf 4 (1).

The firm of Friedrich Krupp AG was approached to study the feasibility of mounting the 12.8cm L/55 gun on the chassis of the Tiger II (or Tiger H3 as it was known at the time) and Henschel & Sohn of Kassel was contracted to develop the suspension, hull and superstructure. The guns, later referred to as the Pak 44, were to be taken unchanged from the Pzkpfw Maus project which had been under development for almost a year.

April 1943. The designers at Henschel had been tasked with examining two designs, one with the engine installed in the hull rear behind the armoured superstructure and a second with a forward-mounted engine. These were referred to as Design A and Design B respectively in later correspondence. In an HWA report dated 12 April it was argued that Design B presented a number of difficulties which were not present with Design A and furthermore the latter provided a degree of uniformity of parts with the Tiger II design although the hull of the new design was some 300mm longer. Henschel's suggestion that the hull and superstructure be built as two separate components to ease maintenance was rejected on the grounds that this would affect waterproofing. It was agreed that a full size wooden version of Design A be presented to Hitler for his approval.

In the HWA report the project was now referred to as both 12.8cm Panzerjäger ausf Tiger H2 and Tiger-Jäger, probably for the first time, and Wa Prüf 6 assumed responsibility for the programme (2).

.........................text continued on page 51

Notes

1. Under the auspices of the Heereswaffenamt (HWA), the army's ordnance office, the Amtsgruppe für Entwicklung und Prüfung was divided into twelve different sections and Wa Prüf 4 was responsible for matters related to the artillery arm. In July 1944 the HWA underwent a major reorganisation but that need not concern us here.

2. Wa Prüf 6 was concerned with all aspects of motor vehicle and tractor development.

Partially completed Jagdtigers, including Fgst Nr.305053 nearest the camera, photographed in Halle VII at the Nibelungenwerk. Note the opened commander's hatch, the three sets of track hangers on the superstructure side and the gun travel lock. This photograph was probably taken in January 1945.

Above: Fgst Nr.305001, the first production series Jagdtiger built with the Porsche suspension and roadwheels (1), photographed during the spring of 1944 when it was undergoing trials. This vehicle, and Fgst Nr.305002, were fitted with the Gg 24/800/300 double-link tracks (2) and the Henschel-designed 18-tooth drive sprocket (3) and rear idler (4). A photograph of Fgst Nr.305002 is reproduced on page 11. In addition, both had lengthened Tiger II trackguards (5), identified by the arrangement of the bolts, which were replaced by purpose-built versions from Fgst Nr.305003 onwards. The cast ribs on the gun mantlet (6), also shown inset (7), are just visible here. Fgst Nr.305001 was later upgraded with track hangers, tool racks on the hull sides, a gun travel lock and other minor modifications and a photograph can be seen on page 12. Below: Fgst Nr.305003 photographed at the Nibelungenwerk in mid-1944 when it was fitted with the Kgs 64/640/130 tracks (8) in an effort to improve the stability of the Porsche suspension. The tracks on the superstructure side are the Gg 24/800/300 model (9) with the inner guide horns cut away, dating this image to at least July. The application of Zimmerit, where it reached the level of the lower set of track hangers on the superstructure sides (10), began with this vehicle and continued to Fgst Nr.305008. Note the ring fixed to the Kugelblende machine-gun mount (11) which may have been intended to hold a waterproof cover. The Porsche-designed rear idler (12) is clearly visible in this image as is the lead (13) for the Bosch headlight (14) and the cast ribs of the gun mantlet (15).

Above: A rear view of Fgst Nr.305004 showing the Zimmerit anti-magnetic mine coating. Note that the paste has also been comprehensively applied to the trackguards (1), fenders (2) and exhaust covers (3). The latter were fitted to this vehicle and Fgst Nr.305003, shown on the previous page, only. The bare patch to the right of the exhaust cover (4) is where the wooden jack block was carried and the brackets for the 20-ton jack (5) are visible on the lower hull. The spare tracks on the superstructure side (6) are the Gg 24/800/300 version with the inner guide horns cut away.

.........................text continued from page 49

August 1943. Hitler personally ordered that the new design but to be referred to as a Sturmgeschütz and not Pak selbstfahrlafette (Sfl.), or self-propelled anti-tank gun. In the same month the name Jagd-tiger was recorded in the Kriegstagebuch, or war diary, of the General der Artillerie and Jagdtiger was mentioned in an instruction from Chef des Generalstabs to the Generalinspekteur der Panzertruppen (1).

September 1943. Considering the delays being experienced with the devleopment of the Pak 44, representatives of Wa Prüf 6 requested that Henschel and Krupp study the feasibility of utilising the 12.8cm L/70 gun, a somewhat longer weapon but by the following month this idea had been rejected.

October 1944. At Arys, in East Prussia, the full-size wooden model was presented to Hitler who insisted that production be commenced as soon as possible. The Henschel-designed suspension, hull and superstructure were adopted and the decision on fitting the 12.8cm L/55 gun was made. Final assembly was to take place at the Nibelungenwerk at Sankt Valentin, south-east of Linz in northern Austria.

December 1943. The planned manufacturing schedule submitted by the HWA would have seen the delivery of the first Jagdtiger in this month but the demand that the Nibelungenwerk increase production of the Pzkpfw IV meant that none of the heavy tank destroyers were assembled.

January 1944. Professor Ferdinand Porsche convinced Hitler to incorporate his suspension system into the Jagdtiger production programme. The Porsche design featured four units on either side of the hull each made up of two 700mm roadwheels fitted to externally-mounted bogies and this system had enjoyed some success with the Ferdinand/Elefant series of tank destroyers. Each unit could be removed individually for maintenance unlike the Henschel arrangement which was made up of interleavered wheels using the same torsion bar system employed with the Tiger and Panther designs. In addition, Porsche's suspension reduced the overall weight by 1,200 kilograms, increased the available space in the fighting compartment and, as it required 450 fewer man hours to produce, was significantly cheaper (2).

The incorporation of the Porsche design added further to production delays and none of the three vehicles which were planned to have been completed in January were delivered.

February 1944. The first two Jagdtigers left the assembly lines. Fgst Nr.305001 was fitted with the Porsche-designed suspension with 700mm-diameter road

Notes

1. Behind these seemingly petty name changes lurked the inter-branch rivalry and empire building that continued until the last day of the war. The Sturmgeschütz batteries were, at this time at least, under the control of the artillery while self-propelled anti-tank units were administered by the commander of the Panzer troops.

2. Porsche had made a considerable investment in this system which was a feature of the company's unsuccessful bid for the Tiger II programme, the Type 180. Interestingly, Professor Porsche's arguments proved to be largely spurious as production time was not decreased, any redesign of the fighting compartment interior was unrealistic at this stage and the Porsche bogies and roadwheels actually raised the profile of the completed vehicle.

Notes

1. Despite other changes to the Jagdtiger track system and drive sprockets the Verladekette was retained until the end of production, albeit with minor variations, and there is some evidence that a number of vehicles operated with this track only in the last weeks of the war.

wheels while Fgst Nr.305002 was built on the Henschel suspension which utilised the 800mm-diameter road wheels of the Tiger II. Both had the 18-tooth drive sprocket, the Henschel-designed rear idler and the two-part Geländekette, or cross-country track, designated as type Gg 24/800/300. These double-link tracks were originally developed for the Panther II and are identified by the prefix Gg. The first number identified the model, in this case 24, the second indicated the track width and the third referred to the pitch, both measured in millimetres. The exact meaning of the prefix is unknown but it has been plausibly suggested that it stands for Guss-stahl, or cast iron bridge link, and geschmiedet or forged connecting link. Again, these were a feature of the Tiger II and it should be noted that with this type of track the links could only be added or removed in pairs. At 800mm this track was far too wide for railway transport and a narrow 660mm version, referred to as Verladekette Gg 24/660/300, was issued (1).

A Bosch headlight was fitted to the hull front and the electrical lead ran vertically in a straight line from the headlight over the edge of the glacis plate. Fgst Nr.305001 was re-worked, probably in December 1944, with the modifications that had been introduced up to that time and was almost certainly the only Jagdtiger built with the Porsche suspension and fitted with the 9-tooth drive sprocket.

March 1944. The HWA production target of four completed vehicles was not met and no deliveries were made in this month.

April 1944. The HWA production target of eight completed vehicles was not met and no deliveries were made in this month.

May 1944. The HWA production target of fifteen completed vehicles was not met and no deliveries were made in this month. Perhaps acknowledging the reality of the situation Wa Prüf 6 submitted a revised production plan which would come into effect in June.

June 1944. Although the Wa Prüf 6 schedule revised the planned production target from twenty completed vehicles down to just two, only one Jagdtiger, Fgst Nr.305003, was completed in this month. This vehicle was fitted with sheet metal covers for the exhaust mufflers and Zimmerit anti-magnetic mine paste. The latter was applied to all vertical surfaces, including the muffler covers, and superstructure sides to the level of the lowest track hanger. In an effort to rectify the problems that were experienced with the Porsche suspension, Kgs 64/640/130 tracks, which had been used with the Ferdinand/Elefant project, were fitted to Fgst Nr.305003.

But the level of instability was still deemed unacceptable and the decision was made to continue production with the Henschel suspension only. This change could not take place immediately and Jagdtigers

Below: Photographed on the assembly line at the Sankt Valentin Nibelungenwerk, this Jagdtiger has been very helpfully identified as Fgst Nr.305054 and the image dated to January 1945 thanks to the chalk markings on the hull side. It was one of the vehicles allocated to schwere Panzerjäger-Abteilung 512 in the following month. The Henschel road wheels were fitted to the axles (1) that were attached by an arm (2) to a transverse torsion bar (3) which passed through the floor of the hull. The torsion bars for the road wheels on the opposite side were anchored in the small holes (4) between the arms seen here and this meant that the road wheel arrangement was slightly asymmetrical. The 9-tooth drive sprocket (5) is just visible as is the Henschel-designed rear idler (6). Note the three sets of track hangers (7) on the superstructure side. Photographs of this vehicle taken later in the production process suggest that it may have been completed as a Jagdtiger als Befehlspanzer, or command vehicle, fitted with the armoured cover (8) for the antenna insulator of the FuG 8 radio on the superstructure rear.

Above: Production details mentioned in the text. 1. The Gg 24/800/300 double-link tracks fitted to the Porsche suspension. A complex design, the bridging or connecting links (A) were made up of four separate parts. Note also the 18-tooth drive sprocket (B). The vehicle depicted in this image is actually the prototype. 2. The Kgs 73/800/300 track and the 9-tooth drive sprocket, shown here on a Tiger II. Both were introduced into Jagdtiger programme in September 1944 with Fgst Nr.305013 and fitted to the remaining Henschel-suspension vehicles. The bridging link of these tracks (C) consisted of a single cast piece. 3. The narrow Verladekette Gg 24/660/300 transport tracks, here fitted to a Porsche-suspension Jagdtiger with the 18-tooth drive sprocket. 4. The first pattern external gun travel lock here shown on Fgst Nr.305009 which left the assembly line in August 1944. Note also the configuration of the headlight's electrical lead (D) as mentioned in the text. 5. From Fgst Nr.305010, the vehicle shown here, the gun travel lock featured a horizontal brace (E). 6. The third and final pattern of the gun travel lock, introduced with Fgst Nr.305031 in December 1944, had a strengthening bar under the locking mechanism (F) and two small brackets (G) to hold the lock in position when it was lowered.

with the Porsche suspension units were produced until September. Photographs of Fgst Nr.305003 show that it was fitted with the Porsche-designed rear idler, unlike Fgst Nr.305001, and this appears to have remained in use until the completion of Fgst Nr.305012.

July 1944. Production delays continued to dog the project and from a planned five Jagdtigers just two vehicles were completed in this month, Fgst Nr.305004 and Fgst Nr.305005, and the sheet metal exhaust covers were fitted to the former and then dropped from production.

From Fgst Nr.305004 the inner row of guide horns on the Gg 24/800/300 track links were cut away. Trials had shown that they tended to jam when the vehicle was in motion and in any case they were not engaged by the Porsche wheels.

Also beginning with Fgst Nr.305004, small tie-down loops were welded along on the upper edge of the fighting compartment and an additional handle was welded to the outside of the driver's and radio operator's hatches.

Beginning in July four sets of track hangers were welded to the superstructure sides. The armour of Fgst Nr.305005 was found to be defective and it was never assigned to an operational unit, probably remaining at Scheiss-Schule Putlos and used for gunnery training.

August 1944. An external travel lock for the 12.8cm gun was introduced. The initial version consisted of an inverted V-shape which was anchored by brackets to the hull front and this type was probably fitted from Fgst Nr.305006 to Fgst Nr.305009. From Fgst Nr.305010 a horizontal brace was added and photographs suggest that this type was also retrofitted to some earlier production vehicles including Fgst Nr.305001. Fgst Nr.305003 may not have ever been fitted with a gun travel lock and Fgst Nr.305004 certainly was not.

The cast gun mantlet of Fgst Nr.305010 featured a number of ribs running back from the centre section and this type was also fitted to Fgst Nr.305001, Fgst Nr.305002 and Fgst Nr.305003.

From Fgst Nr.305006, a small wedge was cut into the frontal armour of the superstructure were it met the hull side. Also, beginning with this vehicle, the electrical lead from the Bosch headlight was bent towards the right side of the hull and again above the Kugelblende machine gun mount towards the top edge of the glacis.

On 19 August 1944 the Heereswaffenamt advised the assembly plants that the practice of units in the field creating their own patterns of camouflage would halt immediately and that a more-or-less standard scheme would be applied to vehicles before they left the factories. The base colour of Dunkelgelb RAL 7028 would be retained, over which large patches of Olivgrün RAL 6003 and Rotbraun RAL 8017 were to be painted (1). Photographic evidence would suggest that none of the August-production Jagdtigers were painted in a factory-applied scheme and examples are shown in the Camouflage & Markings section of this book.

Although the Wa Prüf 6 schedule called for ten Jagdtigers to be assembled in August, a significant decrease from the original thirty-five demanded by the HWA, just five vehicles were delivered.

September 1944. An order of 9 September 1944 directed that the application of Zimmerit was to cease immediately. Jagdtiger Fgst Nr.305010 was the last vehicle to receive a coat of the anti-magnetic mine paste. As mentioned above, a modified external gun travel lock fitted with a horizontal brace was introduced with this vehicle.

Fgst Nr.305011 was probably the first production vehicle to leave the assembly plant with a factory-applied camouflage scheme and it is depicted on page 18 of the Camouflage & Markings section.

The last production vehicle with the Porsche suspension, Fgst Nr.305012, left the assembly line (2).

From Fgst Nr.305013, built with the Henschel suspension, the edges of the armour plate in front of the periscopes of the fighting compartment were bevelled to allow a better view and an anti-aircraft machine-gun mount, referred to as the Fliegerdrehstütze 36, was fitted to the engine deck. The base for the mount was initially bolted in place on the right side of the engine access door between the hinges but it was later moved to a more convenient position next to the coolant filler cap.

Also from Fgst Nr.305013, the Gg 24/800/300 tracks, which had been fitted to all the Porsche suspension vehicles, were replaced by the Kgs 73/800/300 model. This was also a double-link design and may have been a revised version of the Gg 26/800/300 track which was introduced into production of the Tiger II in May 1944. As only the bridging link of the new track system engaged the drive sprocket the 18-tooth sprocket was replaced by a 9-tooth model. (3).

Beginning with Fgst Nr.305017, an armoured cover was fitted over the deep-fording air intake on the engine deck between the oil and coolant filler caps.

In this month, possibly beginning with Fgst Nr.305020, the track hangers were repositioned with the front set being moved slightly forward. Interestingly, Fgst Nr.305020, which survived the war (4), appears to have been assembled without the second handle on the driver's and radio operator's hatches which were introduced into production in the previous July.

In September 1944 production finally began to match expectations with ten of the projected twelve Jagdtigers being completed.

October 1944. A heavy bombing raid severely curtailed production and just five vehicles of the projected fifteen were assembled in this month.

On 31 October 1944 a directive issued by the Heeres-Abnahmestelle II(K) stipulated that the use of Dunkelgelb RAL 7028 as a base colour for armoured vehicles was to cease. They were from now on to be painted in a camouflage scheme made up of patches of colour referred to as Braun, Grün and Gelb over the basic red oxide primer. It may be that this order was merely the formalisation of a practice that had begun, at least at the Henschel plant, in the previous September at around the time Zimmerit was discontinued (5).

Although this image is of mediocre quality I have included it here as it shows Fgst Nr. 305011, the first series production Jagdtiger to leave the Nibelungenwerk after the application of Zimmerit was discontinued. Although it has been suggested that Fgst Nr.305049 was completed with some Porsche parts, the vehicle shown here is officially the last Jagdtiger fitted with the Porsche road wheels, 18-tooth drive sprocket and Porsche-designed rear idler.

Notes

1. The RAL colour system and the colours used by the German army are discussed in detail in the tenth book in this series *TankCraft 10: Tiger I German Army Heavy Tank, Southern Front North Africa, Sicily and Italy.*
2. Although generally accepted this may not be strictly correct. See Fgst Nr.305049 assembled in December 1944.
3. The Kgs prefix indicated Schnelläufige Ketten für Kraftfahrzeuge (K), or high-speed track for motor vehicles, Stahlguss aller Legierungen (s), for cast steel alloy, with schwimmende Bolzen (s) referring to the type of track pin. It appears to have been applied with all track types, except the earlier double-link designs mentioned here, and its use with the Kgs 73/800/300 and Kgs 73/800/152 models may have been a later attempt to simplify the system.
4. This vehicle was for many years on display at the US Army Ordnance Museum and, at the time of writing, is undergoing restoration at Fort Benning.
5. The surviving copy of this order was addressed to the firm of Henschel, the manufacturer of the Tiger, but it would seem safe to assume that the directive applied to all manufacturers.

Below: Fgst Nr.305058 photographed in April 1945. Note the wedge cut from the lower edge of the front plate of the superstructure where it meets the hull (1). The gun travel lock (2) is the third type, with a reinforcing bar (3) below the locking mechanism (4), and is seen here in the folded position. Note the three sets of track hangers (5) on the superstructure side, a number of which have broken off (6), and the small tie-down loops (7). Note also the two handles of the forward hatches (8). This vehicle is fitted with Gg 24/800/300 tracks (9) which were introduced into production in September 1944 with Fgst Nr.30513, the first Jagdtiger built with the Henschel suspension. The absence of Pilze, or sockets, on the superstructure would suggest that this vehicle was assembled before February 1945.

November 1944. Beginning with Fgst Nr.305026, the 20-ton jack and wooden jack block, which had been fitted at the rear hull below the armoured exhaust covers, were dropped from production.

On 29 November 1944 the HWA advised vehicle manufacturers that a new system was to be introduced where the armour components were to be painted in a base coat of Dunkelgrün RAL 6003 before they were delivered to the assembly plants. Once assembly was completed the tank would be camouflaged by the addition of Rotbraun RAL 8017 and Dunkelgelb RAL 7028 painted in sharp contours. This scheme was not to come into effect until 1 March 1945, probably in anticipation of the approaching winter months.

There is some debate over the number of Jagdtigers completed in this month with figures of six or ten being given. I personally believe the former to be correct with the remaining four delivered in December. In any case, it was far short of the planned total of twenty.

December 1944. From Fgst.Nr. 305032 a third type of gun travel lock, which featured a strengthening bar under the locking mechanism, was introduced. At the same time a third set of track hangers were welded to the superstructure sides at the centre and the number of fuel vent lines on the engine deck was increased from two to four. Interestingly, a number of authoritative sources suggest that Fgst Nr.305049, assembled towards the end of the month, was completed with Porsche suspension parts as mentioned on page 13.

On 20 December 1944 the HWA advised that, contrary to the 29 November directive, the painting of armour components in Dunkelgrün RAL 6003 was to begin immediately.

Twenty vehicles left the assembly plant in this month, the highest monthly production total achieved, but still short of the projected twenty-five.

January 1945. From Fgst Nr.305052 a number of vehicles were fitted with experimental shrapnel covers over the air intakes on the rear deck. Although it has been suggested that as many as ten Jagdtigers were fitted with these armour plates it should be remembered that just a single Tiger II, Fgst Nr.280404, received this modification and it is unlikely that this innovation effected more than a handful of Jagdtigers. Beginning with Fgst Nr.305058, a handle was welded above the rear access door of the superstructure.

The HWA issued a directive at the end of the month which called for the production of 100 Jagdtigers by May 1945 (1) at which time the Nibelungenwerk would switch to production of the Tiger II. As an amendment to the 20 December 1944 order regarding the use of RAL 6003 Dunkelgrün as a vehicle base coat, a

Notes

1. Ten in January, which was an accomplished fact by the time the report was written, forty projected for February, twenty-five in March and a further twenty-five in April.

directive of 23 January 1945 advised manufacturers that existing paint stocks could be used until 30 May 1945, provided that they were not thinned. It was anticipated that the new camouflage would be in general use by June 1945.

February 1945. Beginning with Fgst Nr. 305062, four sockets or Pilzen were welded to the superstructure on either side and at the front and rear to accommodate the 2-ton Befehlskran.

In February 1945 Hitler ordered that every effort was to be made to increase Jagdtiger production, including the use of weapons other than the 12.8cm gun, specifically the 8.8cm L/71.

Although the planned completion figure of forty vehicles, as mentioned above, was revised to thirty, just thirteen Jagdtigers were completed in February.

March 1945. A devastating air raid on 23 March further delayed production and just seven vehicles were accepted by the HWA in this month. Of this total, just three or perhaps four were new Jagdtigers with the remainder being repaired vehicles.

As mentioned above, a new track design, the Kgs 73/800/152, was supposed to have been introduced into the production run of both the Tiger II and Jagdtiger in March but it is unlikely that any were fitted. Photographs taken in the last weeks of the war of abandoned Jagdtigers fitted with narrow tracks almost certainly depict the Verladekette Gg 24/660/300.

April 1945. In the previous month a new type of front fender, reinforced with a raised vertical rib, was introduced into the Tiger II production run. This feature may have been incorporated into the Jagdtiger programme at about the same time and was certainly fitted to Fgst Nr.305083, the April production vehicle which is today on display at the Kubinka Museum. This Jagdtiger was also fitted with extra sockets, situated at each corner of the superstructure roof, and a heavy towing bracket which was bolted to the rear hull and anchored on either side to the armoured exhaust covers. A bracket for a large two-handed saw was fitted on the right side of the hull towards the rear, positioned inside the loop created by the towing cable. This vehicle, and its modifications, may represent the final production version of the Jagdtiger.

An intriguing report compiled for the Inspekteur der Panzertruppen and dated 29 April 1945 clearly states that deliveries made since 4 April had consisted of four 'Jagdtiger 12.8' to s.Pz.Jg.Abt 653 and approximately four 'Jagdtiger 8.8'. Unfortunately no unit allocation is given for the latter. It goes on to mention that a further seventeen 'Jagdtiger 8.8' were expected to be available in May (1).

May 1945. Accurate production figures are not available for this month and any vehicles that were assembled or repaired were destroyed before the Nibelungenwerk was abandoned on 4 May.

Notes

1. The question of 8.8cm-armed Jagdtigers is examined on page 15 and mentioned in the unit histories.

Below: Photographed during the spring of 1945 in the town of Bad Driburg, east of Paderborn, this Jagdtiger is almost certainly one of the vehicles assigned to 3.Kompanie, schwere Panzerjäger-Abteilung 512 in the last weeks of the war. Note the Pilze, or sockets, welded to the superstructure side (1) and above the gun mantlet (2), a production feature introduced in February 1945. Note also that the radio operator's hatch, and the driver's hatch which is visible in other photographs of this vehicle, has only one handle (3) although a second handle was supposed to have been fitted from July 1944. This can also be seen on Fgst Nr.305020, a September 1944 production model and it may be that this step was omitted on some vehicles due to time constraints or that old stocks were used. Many contemporary accounts mention that tank crews were drafted in to help with assembly at the Nibelungenwerk and this go some way towards explaining why the two forward sections of both trackguards (4 and 5) are fitted backwards, albeit neatly. This vehicle is also shown in the Camouflage & Markings section on page 24.

Both images reproduced here depict the vehicle shown on the previous page. Points to note include: 1. The cast gun mantlet with lifting bracket (2). 3. Radio operator's hatch with single handle as mentioned previously. The section containing both hatches was bolted into the hull roof and has here been forced upward by an internal explosion. 4. Part of the hull trackguard. Note how these are attached to the hull side. The trackguards are sometimes, incorrectly, referred to as Schürzen but those fitted to the Jagdtiger and Tiger II were not intended to provide any additional armour protection but were merely designed to prevent excesses of mud and dirt coating the hull. 5. The Kugelblende machine-gun mount. 6. The third pattern gun travel lock distinguished by the horizontal brace (7) with strengthening bar (8), situated below the locking mechanism (9). The latter was hinged on the left side, not visible here, and could be secured by lowering the handle (9). Note the small brackets on the glacis (10) which held the frame when it was folded flat. From September 1944 these vehicles were fitted with the Fliegerdrehstütze 36 anti-aircraft mount, shown below. The stand (11) could be removed from the base (12) which was bolted to the engine access door (13). The base was originally situated on the opposite side of the door, where the hinged hook (14) can be seen here, but was later moved to a point behind the access door next to the coolant filler cap, at least by Fgst Nr.305083. The position of the base shown here may indicate that this was an interim modification or that it was retro-fitted. The stand featured an extension, shown inset (15), and a bracket (16) to which the MG 42 was attached. This photograph also provides a good view of the Henschel-designed rear idler (17) and the Kgs 73/800/300 track (18) with a badly worn guide horn.

JAGDTIGER PRODUCTION DETAILS

Although the modifications made during the Jagdtiger production run are discussed in some detail on the previous pages, and additional information is included in the tables on pages 11 to 15, this chart is intended to provide readers with enough information to identify individual Jagdtigers and to this end I have included only the most obvious and easily recognisable features. It should be remembered that vehicles returned for repairs were often upgraded with the most recent modifications and the details mentioned here could have changed at some time. Fgst Nr.305001 is a prime example as can be seen by comparing the photographs reproduced on page 50, when it was first accepted by the

Fgst Nr.	Porsche Suspension	Henschel Suspension	18-tooth Drive Sprocket	9-tooth Drive Sprocket	Porsche Rear Idler	Henschel Rear Idler	Gg 24/800/300 Track	Kgs 64/640/130 Track	Kgs 73/800/300 Track	Zimmerit *To Lowest Track Hanger*	Zimmerit *To Upper Track Hanger*	Sheet Metal Exhaust Covers	20-Ton Jack and Wooden Block	Tie-Down Loops on Superstructure Sides	Wedge in Frontal Armour	Gun Travel Lock 1	Gun Travel Lock 2	Gun Travel Lock 3	Track Hangers, 4 sets	Track Hangers, 6 sets	Additional Handle on Forward Hatches	Armoured Cover for Deep-Fording Air Intake
305001																						
305002																						
305003																						
305004																						
305005																						
305006																						
305007																						
305008																						
305009																						
305010																						
305011																						
305012																						
305013																						
305014																						
305015																						
305016																						
305017																						
305018																						
305019																						
305020																						
305021																						
305022																						
305023																						
305024																						
305025																						
305026																						
305027																						
305028																						
305029																						
305030																						
305031																						
305032																						
305033																						
305034																						
305035																						
305036																						
305037																						
305038																						
305039																						
305040																						
305041																						
305042																						
305043																						
305044																						
305045																						
305046																						

HWA, and page 12, taken some months later. In fact, photographic evidence, where it exists, is the only way we can be sure of which modifications were made to certain vehicles. Similarly the information shown here for Jagdtigers built from Fgst Nr.305085 is highly speculative and it is possible that most of these vehicles were never fully assembled. The inclusion of Jagdtigers up to Fgst Nr.305092 is a purely arbitrary decision on my part as explained earlier. Note also that I have not made mention of the main armament as in all verified cases this was the 12.8cm L/55 gun and the possible installation of the 8.8cm is a matter of debate.

Fgst Nr.	Porsche Suspension	Henschel Suspension	18-tooth Drive Sprocket	9-tooth Drive Sprocket	Porsche Rear Idler	Henschel Rear Idler	Gg 24/800/300 Track	Kgs 64/640/130 Track	Kgs 73/800/300 Track	Zimmerit *To Lowest Track Hanger*	Zimmerit *To Upper Track Hanger*	Sheet Metal Exhaust Covers	20-Ton Jack and Wooden Block	Tie-Down Loops on Superstructure Sides	Wedge in Frontal Armour	Gun Travel Lock 1	Gun Travel Lock 2	Gun Travel Lock 3	Track Hangers, 4 sets	Track Hangers, 6 sets	Additional Handle on Forward Hatches	Armoured Cover for Deep-Fording Air Intake
305047																						
305048																						
305049																						
305050																						
305051																						
305052																						
305053																						
305054																						
305055																						
305056																						
305057																						
305058																						
305059																						
305060																						
305061																						
305062																						
305063																						
305064																						
305065																						
305066																						
305067																						
305068																						
305069																						
305070																						
305071																						
305072																						
305073																						
305074																						
305075																						
305076																						
305077																						
305078																						
305079																						
305080																						
305081																						
305082																						
305083																						
305084																						
305085																						
305086																						
305087																						
305088																						
305089																						
305090																						
305091																						
305092																						

SCHWERE PANZERJÄGER-ABTEILUNG, LATE 1944

All formations of the German Army were organised according to detailed tables of establishment referred to as Kriegsstärkenachweisung (KstN) which showed the official composition of a unit, listing the exact number of personnel and type of equipment from vehicles to small arms. The tables were numbered and dated, the latter being all-important as some numbers were duplicated. The KstN were accompanied by Kriegsausrüstungsnachweisung (KAN) which listed minor items in detail such as tools and fuel containers. Unfortunately, the KstN and KAN relating to the formation of the Jagdtiger battalions have not survived but there are a number of official documents and orders which provide several clues as to how those units were organised. The most detailed is a Gliederung, or organisational chart, compiled for schwere Panzerjäger-Abteilung (s.PzJg.Abt) 653 which shows the battalion's composition, down to company level, and its authourised armoured vehicles. This chart must have been completed at some time after April 1944 when Major Grillenberger took command. Another indicator is the

........text continued on facing page

Stab,
schwere Panzerjäger-Abteilung

Major Rudolf Grillenberger as battalion commander with Oberleutnant Kurt Scherer as battalion adjutant. No vehicles are listed on the Gliederung but a report submitted in February 1945 mentions that the battalion had in total thirty-seven staff cars and sixteen motorcycles. KstN 1106 dated 1 February 1944 for Stab und Stabskompanie einer Panzerjägerabteilung (gem.) gives two motorcycles, two VW light cars and two staff cars and this formation may have been similar.

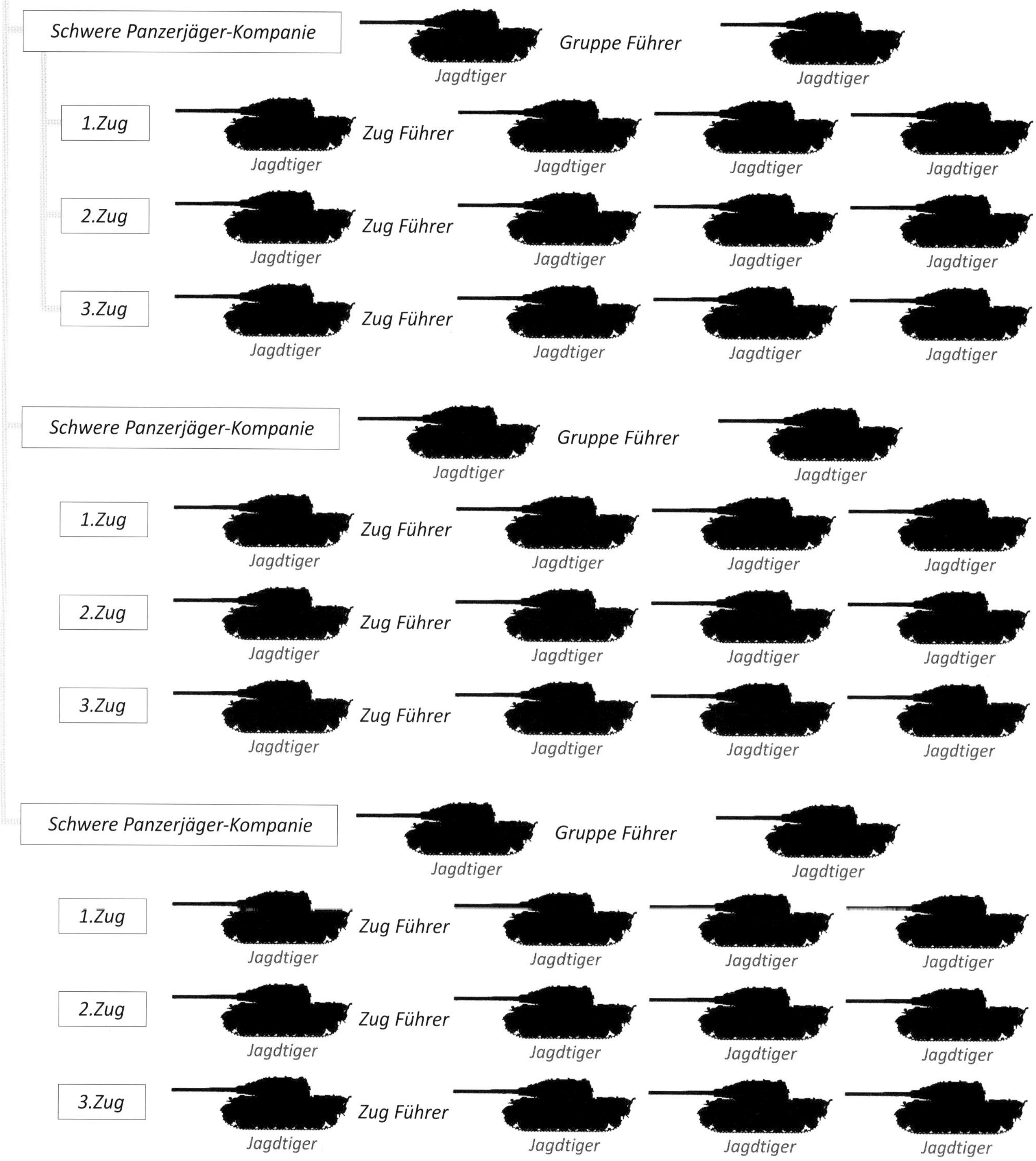

February 1945 order issued for the creation of s.PzJg.Kompanie 614 , built around 2.Kompanie, s.PzJg.Abt 653, which stipulates the use of an amended version of KstN 1148c. This was the establishment used to form the schwere Panzerjäger companies of the battalion in mid-1943 when it was equipped with the Elefant tank destroyer and the possible relevance of the February order is discussed in the unit histories. It is entirely possible that the 1943 establishments were retained with amendments and additions to the support elements, taking into account the frei Gliederung principles that had been gradually introduced from early 1944. It should also be remembered that the formation of s.PzJg.Abt 653, and the beginning of series production of the Jagdtiger, roughly coincided with the creation of the schwere Panzerjäger-Abteilungen equipped with the Jagdpanther and it is likely that at least parts of the KstN used for those battalions were adapted for the Jagdtiger units. Shown below is the organisation of s.PzJg.Abt 653 based on the Gliederung mentioned above and this should be compared with the actual configurations of s.PzJg.Abt 653 and s.PzJg.Abt 512 which are examined on the following pages.

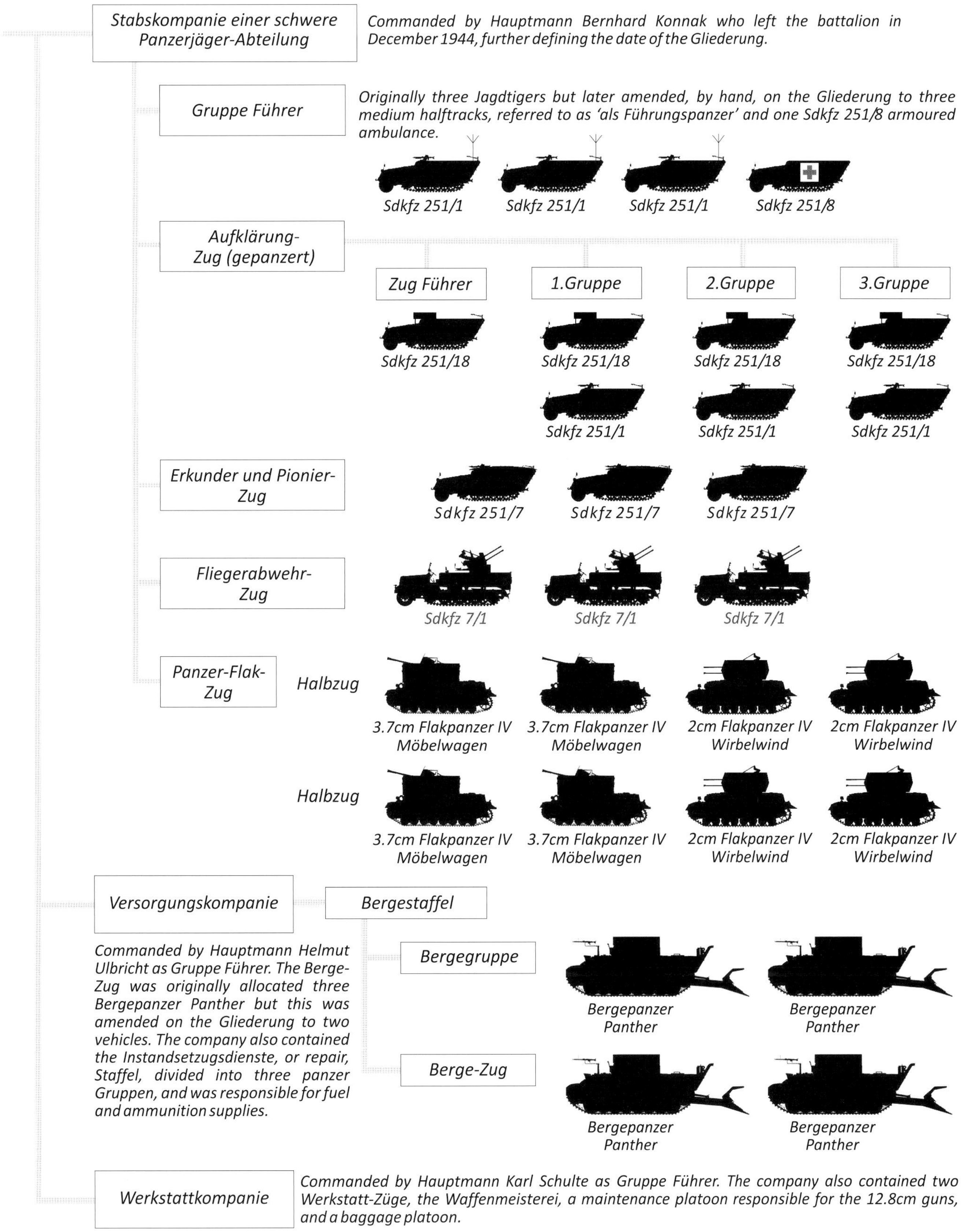

SCHWERE PANZERJÄGER-ABTEILUNG 653, SPRING 1945

Any attempt at constructing a detailed organisational table for this formation is made difficult by the fact that elements of the battalion were sent to the front at different times and were usually attached to Kampfgruppen as the situation dictated. In addition, the official documents that have survived the war are often ambiguous and sometimes contradictory. For example, the very comprehensive report of 23 January 1945 states that the battalion had finally achieved its authorised strength of forty-two Jagdtigers, and this number had in fact been delivered, but this ignores the vehicle that was lost in combat earlier that month. The Fliegerabwehr-Zug, with its three Sdkfz 7/1 halftracks, is also mentioned in the January report, and touched on briefly in at least one veteran's account, but seems to have disappeared by March 1945. In the second week of that month, in a status report submitted by the battalion, four Flakpanzer IV Ostwind are mentioned and the accuracy of this statement has to be questioned. These vehicles were almost certainly the four 3.7cm-armed Möbelwagen that had been on hand for some time. Although the battalion was authorised five Bergepanther recovery tanks, by March just three were available and one of these were probably assigned to each company. The three-digit numbering system shown here was generally used throughout the battalion but the numbers varied in style and size. Examples, and one known anomaly, are shown in the Camouflage & Markings section. The lighter colours indicate the existence of photographs of those vehicles.

Stab, schwere Panzerjäger-Abteilung

Kommandeur des Bataillons, Major Rudolf Grillenberger to April 1945 then Major Rolf Fromme

Stabs-Kompanie — *Gruppe Führer, Hauptmann Bernard Konnak*

3 x Sdkfz 251

- Aufklärung-Zug (gepanzert): 7 x Sdkfz 251
- Erkunder und Pionier-Zug: 3 x Sdkfz 251
- Fliegerabwehr-Zug: 3 x Sdkfz 7/1
- Panzer-Flak-Zug: 8 x FlakPanzer IV

Not listed here but subordinated to the Stabs-Kompanie are non-combat support elements including the Kfz.Instandsetzungsstaffel, Gefechtstross, Sanitätstrupp, Staffel für Verwaltung und Nachschub, Verpflegungstross and Gepäcktross.

Versorgungs-Kompanie

Werkstatt-Kompanie

1.Kompanie: 101, 102 — *Gruppe Führer, Oberleutnant Werner Haberland*

- 1.Zug — *Zug Führer, Leutnant Hans Knippenberg*: 115, 112, 113, 114
- 2.Zug — *Zug Führer, Oberfeldwebel Gustav Koss*: 121, 122, 123, 124
- 3.Zug — *Zug Führer, Oberfeldwebel Ferdinand Kinnberger*: 131, 132, 133, 134

2.Kompanie: 201, 202 — *Gruppe Führer, Oberleutnant Robert Weisenfarth*

- 1.Zug — *Zug Führer, Leutnant Rudolf Braun*: 211, 212, 213, 214
- 2.Zug — *Zug Führer, Leutnant Feineisen*: 221, 222, 223, 224
- 3.Zug — *Zug Führer, Stabsfeldwebel Ernst Geile*: 231, 232, 233, 234

3.Kompanie: 301, 302 — *Gruppe Führer, Oberleutnant Franz Kretschmer*

- 1.Zug — *Zug Führer, Oberfeldwebel Emil Issler*: 311, 312, 313, 314
- 2.Zug — *Zug Führer, Oberfeldwebel Fritz Schwarz*: 321, 322, 323, 324
- 3.Zug — *Zug Führer, Leutnant Kaspar Göggerle*: 331, 332, 333, 334

SCHWERE PANZERJÄGER-ABTEILUNG 512, SPRING 1945

Not shown below are the lighter vehicles allocated to the battalion such as twelve Sdkfz 2 Kettenkrad halftrack-motorcycles, twenty-five light lorries, three heavy recovery halftracks and assorted cars. An order issued on 30 March 1945 called for the formation of a mixed Panzer-Flak-Zug, as shown on page 61, which was to be partly ready by 3 April, but this almost certainly never materialised.

Stab, schwere Panzerjäger-Abteilung

Versorgungs-Kompanie

Werkstatt-Kompanie

Kommandeur des Bataillons, Major Walther Scherf

1.Kompanie – X1 – *Gruppe Führer, Oberleutnant Albert Ernst (later promoted to Hauptmann)*

- *1.Zug* – *Zug Führer, Oberfähnrich Heinz Rondorf (possibly promoted to Leutnant by this time)* – X2, X3, X4
- *2.Zug* – *Zug Führer, Leutnant Kubelka* – X5, X6, X7
- *3.Zug* – *Zug Führer, Leutnant Sepp Tarlach* – X8, X9, X10

The Jagdtigers of 1.Kompanie were numbered as shown above with the lighter-coloured numbers representing those that have been confirmed by photographic evidence. Many accounts suggest that there was at least a plan to use a similar numbering system, employing Y and Z, for the remaining companies but this was almost certainly never implemented, despite the existence of photographs which show a Jagdtiger simply marked Y that was disabled near Paderborn in early April 1945. In most surviving records and veterans' accounts the Jagdtigers of 2.Kompanie and 3.Kompanie are referred to by either the commander's name or using the official three-digit numbering system. The commander of 2.Kompanie, Oberleutnant Carius, was interviewed many times after the war and had no recollection of the matter. He also stated that, to his knowledge, no photographs had survived the war and any cameras his men possessed were confiscated when they surrendered.

2.Kompanie – *Gruppe Führer, Oberleutnant Otto Carius*

- *1.Zug* – *Zug Führer, Leutnant Hans-Jürgen Hass*
- *2.Zug* – *Zug Führer, Leutnant Friedrich Riess*
- *3.Zug* – *Zug Führer, Leutnant von Salisch*

3.Kompanie – *Gruppe Führer, Oberleutnant Werner Schrader*

- *1.Zug*
- *2.Zug* – *Zug Führer, Feldwebel Otto Arnold* – 321, 322
- *3.Zug*

The total number of eight Jagdtigers, which admittedly disagrees with most accounts, is based on reported losses in the fighting between Sennelager and Sankt Andreasberg as explained in the unit histories. The company's 3.Zug, with three Jagdtigers, is reported to have been destroyed on 15/16 April 1945 in an Allied bombing raid but the allocation of vehicles to the other platoons, as shown here, is purely speculative on my part. Although vehicle numbers are mentioned in some records, such as Jagdtiger 321 of Feldwebel Arnold and 322 commanded by Feldwebel Becker, these may not have been displayed on the vehicles and inconsistencies, for example Y, 55 and 11, are confirmed by photographic evidence.

Dragon Models Ltd
B1-10/F., 603-609 Castle Peak Rd,
Kong Nam Industrial Building,
Tsuen Wan, N. T., Hong Kong
www.dragon-models.com

Tamiya Inc
Shizuoka City, Japan
www.tamiya.com

Italeri S.p.A.
via Pradazzo 6/b,
40012 Calderara di Reno, Bologna, Italy
Www.italeri.com

Hasegawa
3-1-2 Yagusa, Yaizu 425-8711,
Shizuoka City, Japan
www.hasegawa-model.co.jp

Trumpeter/Hobby Boss
NanLong Industrial Park,SanXiang,
ZhongShan,GuangDong, P.R. China
www.trumpeter-china.com
www.hobbyboss.com

Academy Plastic Models
521-1, Yonghyeon-dong, Uijeongbu-si,
Gyeonggi-do, Korea
www.academy.co.kr

Sturmpanzer.com
www.sturmpanzer.com
Copies of original documents including KstN, vehicle allocation lists and more.

Hauler
Jan Sobotka,
Moravská 38, 620 00 Brno,
Czech Republic
www.hauler.cz

Voyager
Room 501, No.411 4th Village,
SPC Jinshan District, Shanghai 200540,
P.R.China
www.voyagermodel.com

Griffon Model
Suite 501, Bldg 01, 418 Middle Longpan Rd,
Nanjing, P.R. China
www.griffonmodel.com

Aber
ul. Jalowcowa 15, 40-750 Katowice, Poland
www.aber.net.pl

E.T. Model
I can personally recommend this company for their service.
www.etmodeller.com

Friulmodel
H 8142. Urhida, Nefelejcs u. 2., Hungary
www.friulmodel.hu

Modelkasten
Chiyoda-ku Kanda, Nishiki-Cho 1-7, Tokyo,
Japan
www.modelkasten.com
Very difficult to navigate but worthwhile.

Eduard Model Accessories
Mirova 170, 435 21 Obrnice,
Czech Republic
www.eduard.com

Master Club
www.masterclub.ru
It appears that this firm is closely associated with Armour35, a Russsian mail-order firm.

RB Model
Powstancow Wlkp.29B,
64-360 Zbaszyn,
Poland
www.rbmodel.com

Kaizen
Like many Chinese companies this is hard to pin down to an exact location. The tracks are available from MS Models Japan, a firm for which I can personally vouch.
www.msmodelswebshop.jp

Passion Models
This Japanese company is affiliated with MS Models Japan.
www.passionmodels.jp

Royal Model
Via E. Montale, 19-95030,
Pedara, Italy
www.royalmodel.com

In writing this book I referred to a number of printed works including *Heavy Jagdpanzer, Development Production Operations* by Walter J. Spielberger, Hilary L. Doyle and the late Thomas Jentz, which although somewhat dated is still a valuable resource. I also consulted *Jagdtiger The Most Powerful Armoured Fighting Vehicle of World War II*, published in two volumes, by Andrew Devey and indispensable to any study of the Jagdtiger battalions is Karl-Heinz Münch's *German Heavy Anti-Tank Unit 653 in World War II. Schwere Panzer Der Wehramcht, Von Der 12.8cm Flak Bis Jagdtiger* by Michael Frölich contains a mass of useful information on vehicle production and allocation but is unfortunately only available in German. I also found the *Panzerwrecks* series of books to contain much helpful, and often obscure, material. Information regarding the allocation of the assembled Jagdtigers was gleaned from the original Führungliste which are available from Sturmpanzer.com and their contact details appear above. I would also like to acknowledge the contributors to both the Axis History and Feldgrau online forums, particularly Martin Block, and I must make special mention of Klaus Delventhal, whose 1/16 scale model appears in this book, who very generously shared the fruits of his research. I would also like to thank Jan Zdiarsky from Eduard Model Accessories, Freddie Leung of Dragon Models and the crew at E.T. Models who are always ready to help. As always, I am indebted to Karl Berne, Valeri Polokov and J.Howard Parker for their invaluable assistance with the photographs and period insignia.

Said to have been abandoned on 15 April 1945 near Sankt Andreasberg, this Jadgtiger is almost certainly one of the vehicles allocated to 3.Kompanie, schwere Panzerjäger-Abteilung 512. Other images of the same vehicle show that it was a Befehlspanzer fitted with the armoured cover on the rear of the superstructure to protect the insulator of the FuG 8 radio antenna. Note that it is fitted with the narrow Verladkette transport tracks. This vehicle is also shown in the Camouflage & Markings section on page 25.